Carpentry and Joinery 1

Carpentry and Joinery

1

Third Edition

Brian Porter LCG, FIOC
Formerly of Leeds College of Building

BUTTERWORTH
HEINEMANN

OXFORD   AMSTERDAM   BOSTON   LONDON   NEW YORK   PARIS
SAN DIEGO   SAN FRANCISCO   SINGAPORE   SYDNEY   TOKYO

Butterworth-Heinemann
An imprint of Elsevier Science
Linacre House, Jordan Hill, Oxford OX2 8DP
200 Wheeler Road, Burlington MA 01803

First published in Great Britain 1984
Second edition published by Edward Arnold 1991
Third edition 2002
Reprinted 2003

British Library Cataloguing in Publication Data
Porter, Brian 1938–
Carpentry and joinery. – 3rd ed.
Vol. 1
1. Carpentry and Joinery.
I. Title
694

ISBN 0 7506 5135 0

For information an all Butterworth-Heinemann publications
visit our website at www.bh.com

Design and typesetting by J&L Composition Ltd, Filey, North Yorkshire
Printed and bound in Great Britain by Martins the Printers

Contents

Foreword

The craft of the carpenter and joiner, at least in those areas of the world where there is plentiful supply of timber is as old as history, and this book describes and illustrates for the benefit of students and others who care to read it, the changing techniques that continue to take place as our knowledge of wood and its working develops. Standards controlling the quality of timber, timber based products, workmanship and safe working practices are continually being revised and harmonised to meet, not only our higher standards, but also those of Europe. Improved fastenings and adhesives have revolutionised joining techniques. Development in electrical battery technology has made possible the cordless power tool. These improvements in the field of woodworking are a continual process, and so must be the updating of textbooks to reflect these changes. Brian Porter as a practicing Carpenter and Joiner, and a lecturer in wood trades is familiar with these changes, which have been incorporated into this revised edition, written to help wood trade students in the early stages of their chosen careers understand the techniques and principles involved in the safe and efficient working of timber and timber products. A book maintaining the high standard Brian Porter set himself in his earlier publications, and which provides a wealth of information that will be helpful to all who have an interest in the working of wood.

Reg Rose MCIOB, DMS, DASTE,
FIOC former Assistant Principal,
Leeds College of Building, UK

Preface

I find it difficult to comprehend that after 20 years and two previous editions, this book is still in demand across the world. Its content has of course been periodically updated in keeping with current trends and legislation, but in essence, it remains the same.

This new edition is printed in a similar format to the last one, but as can be seen from the content page similarities end there. Its new design has taken into account the necessary theoretical job knowledge requirements of the modern carpenter and joiner. But, as with all my previous books, still maintaining a highly visual approach to its content.

It would appear to some people that in recent years, many of our traditional hand tool skills have been replaced by the used of portable power tools. To some extent this may be partly true, but I believe that in the majority of cases, the professional carpenter and joiner would regard portable power tools more as an aid to greater productivity, rather than a replacement for traditional hand tools.

Hand tools still play a vital part in our work, and it is for this reason that I have again included a large section devoted to their selection, safe use and application – together with several pieces of ancilliary equipment.

Carpentry and Joinery volume 1, can also be used by students to help them grasp basic underpinning knowledge of many, if not all of our every day work activities. It should be particulary helpful as a basis for acquiring a greater understanding of the activities set out in the new editions of Carpentry and Joinery volumes 2 and 3.

Unlike previous editions, the practical projects now appear within their own section towards the back of the book. To accompany the now well established 'Porterbox' system of containers, two additional projects have been added to this section – a 'saw stool' and 'workbench'.

I feel confident that the readers of this book will find it not only an asset to gaining a greater understanding of our craft, but also as a reference manual for future use.

Brian Porter 2001

Preface to the second edition

The main difference between this book – the first of three new editions – is the overall size and format of the contents compared to the previous first edition (published in 1982). The most important reason for this change is that in the majority of cases text and illustration can now share the same or adjacent page, making reference simpler and the book easier to follow. The most significant change of all is the new section on tool storage: Several practical innovative projects have been included which will allow the reader to make, either as part of his or her coursework or as a separate exercise, a simple, yet practical system of tool storage units and tool holders – aptly called the 'Porterbox' system (original designs were first published in *Woodworker* magazine). Each chapter has been reviewed and revised to suit current changes. For example, this has meant the introduction of new hand tools, replacing or supplementing existing portable powered hand tools, and updating some woodworking machines.

Educational and training establishments seem to be in a constant state of change; college and school based carpentry and joinery courses are no exception. As the time available for formal tuition becomes less, course content, possibly due to demands made by industry and the introduction of new materials together with a knowledge of any associated modern technology they bring with them, seems to be getting greater – making demands for support resource material probably greater than they have ever been.

Distance learning (home study with professional support) can have a very important role to play in the learning process, and it should be pointed out that in some areas of study it is not just an alternative to the more formally structured learning process, but a proven method in its own right.

No matter which study method is chosen by the reader, the type of reading matter used to accompany studies should be easy to read and highly illustrative, and all subjects portrayed throughout this book meet that requirement. I hope therefore that this book is as well read, and used, by students of this most fulfilling of crafts.

Brian Porter
Leeds 1989

Preface to the first edition

This volume is the first of three designed to meet the needs of students engaged on a course of study in carpentry and joinery. Together, the three volumes cover the content of the City and Guilds of London Institute craft certificate course in carpentry and joinery (course number 585).

I have adopted a predominantly pictoral approach to the subject matter and have tried to integrate the discussion of craft theory and associated subjects such as geometry and mensuration so that their interdependence is apparent throughout. However, I have not attempted to offer instruction in sketching, drawing, and perspective techniques (BS 1192), which I think are best left to the individual student's school or college.

Procedures described in the practical sections of the text have been chosen because they follow safe working principles – this is not to say that there are no suitable alternatives, simply that I favour the ones chosen.

Finally, although the main aim of the book is to supplement school or college-based work of a theoretical and practical nature, its presentation is such that it should also prove invaluable to students studying by correspondence course ('distance learning') and to mature students who in earlier years may perhaps have overlooked the all-important basic principles of our craft.

Brian Porter
Leeds 1982

Acknowledgements

I wish to thank:

Reg Rose for proof reading the text, writing the foreword, and allowing me to reproduce many pieces of artwork we shared in previous joint authorship work as listed on the back of this book.

Eric Cannell for editing and contributing material for Chapter 9 (Woodworking Machines).

Peter Kershaw (Managing Director North Yorkshire Timber Co Ltd.) for his help and guidance.

Colleagues and library staff at Leeds College of Building.

I would also like to offer my gratitude to the following people and companies for their help and support by providing me with technical information and in many cases permission to reproduce item of artwork and/or photographs. Without their help many aspects of this work would not have been possible.

Arch Timber Protection (formerly 'Hickson Timber Products Ltd.)
A L Daltons Ltd (Woodworking machines)
American Plywood Association (APA)
Black & Decker
British Gypsum Ltd.
Cape Boards Ltd.
Cape Calsil Systems Ltd.
Council of Forest Industries (COFI)
CSC Forest Products (Sterling) Ltd.
Denford Machine Tools Co Ltd.
DeWALT
Draper (The Tool Company).
English Abrasive & Chemicals Ltd.
Fibre Building Board Organisation
Finnish Plywood International
Fischer Fixing Systems.
Forestor – Forest and Sawmill Equipment (Engineers) Ltd.
Formica Ltd.
Fosroc Ltd.
G. F. Wells Ltd. (Timber Drying Engineers).
Hilti Ltd.
ITW Construction Products (Paslode)
Kiln Services Ltd.
Louisiana-Pacific Europe
Makita UK Ltd.
Mr Stewart J. Kennmar-Glenhill (Imperial College of Science and Technology, London) and David Kerr and Barrie Juniper of the Plant Science Department, Oxford, for contributing Figures 1.3 to 1.6
Mr. John Common (Kiln Services Ltd.)
Neil Tools Ltd.

Nettlefolds (Woodscrews)
Nordic Timber Council
North Yorkshire Timber Co Ltd.
Perstorp Warerite Ltd.
Protim Ltd.
Protimeter PLC
Rabone Chesterman Ltd.
Record Marples (Woodworking Tools) Ltd.
Record Tools Ltd.
Rentokil Ltd.
Robert Bosch Ltd. (Power Tools Division)
Stanley Tools, Stanley Works Ltd.
Stenner of Tiverton Ltd.
The Rawlplug Company Ltd.
Timber Research and Development Association (TRADA), for information gleaned from TRADA
    Wood Information Sheets'.
Trend Machinery & Cutting Tools Ltd.
Wadkin Group of Companies PLC.
Willamette Europe Ltd.
Wolfcraft

Tables 1.2 and 1.4 are extracted from BS 4471
Table 1.3 is extracted from BS 5450
Copies of the complete standards can be obtained from British Standards Institute, 389
Chiswick High Road, London W4 4AL. Copyright is held by the Crown and reproduced with kind
permission of the British Standards Institute.

With kind permission the line drawings in figures 9.1 (WIS 31). 9.3 (WIS 31). 9.7 (WIS 16). 9.25
(WIS 17). 9.26 (WIS 17). 9.27 (WIS 17). 9.30 (WIS 17). and 9.42 (WIS 31). were extracted from
'Health & Safety Executive' (HSE) Woodworking Information Sheets (WIS).
The full information sheets are available from:
HSE Books
PO Box 1999
Sudbury
Suffolk CO10 2WA

Timber

1

As part of their craft expertise, carpenters and joiners should be able to identify common, commercially used timbers and manufactured boards, to the extent that they also become aware of how they (as shown in figure 1.1) will respond to being:

a cut by hand and machine,
b bent,
c subjected to loads,
d nailed and screwed into,
e glued,
f subjected to moisture,
g attacked by fungi,
h attacked by insects,
i subjected to fire,
j treated with preservatives, flame retardants, sealants, etc.,
k in contact with metal.

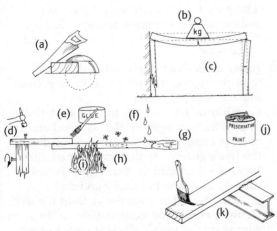

Fig 1.1 Timber may respond differently to these treatments

By and large, behaviour under these conditions will depend on the structural properties of the timber, its working qualities, strength and resistance to fungal decay (durability), insect attack, chemical make-up, and moisture content.

1.1 Growth & structure of a tree

The life of a tree begins very much like that of any other plant – the difference being that, if the seedling survives its early stage of growth to become a sapling (young tree), it may develop into one of the largest plants in the plant kingdom.

The hazards to young trees are many and varied. Animals are responsible for the destruction of many young saplings, but this is often regarded as a natural thinning-out of an otherwise overcrowded forest, thus, allowing the sapling to mature and develop into a tree of natural size and shape. Where thinning has not taken place, trees grow thin and spindly – evidence of this can be seen in any overgrown woodland where trees have had to compete for the daylight necessary for their food production.

With all natural resources that are in constant demand, there comes a time when demand outweighs supply. Fortunately, although trees require 30–100 years or more to mature, it is possible to ensure a continuing supply – provided that land is made available and felling, (cutting down) is strictly controlled. This has meant that varying degrees of conservation have had to be enforced throughout some of the world's largest natural forests and has led to the development of massive man-made forests (forest farming).

1.1.1 Tree components

There are three main parts:

- the root system,
- the stem or trunk,
- the crown.

As can be seen from figure 1.2 these can vary with the type of tree (section 1.2).

a **Root system** – The roots anchor the tree firmly into the ground, these can exceed the radius of the tree – size and spread depends on type and size of tree. The many small root hairs surrounding the root ends, absorb water and minerals, to form sap (see fig. 1.3).

b **Trunk (stem)** – The stem or trunk conducts sap from the roots, stores food, and supports the crown. When the trunk is cut into lengths – they are called logs or boules. Timber is cut from this part of the tree (see fig. 1.3).

c **Crown** – The crown consists of branches, twigs, and foliage (leaves). Branches and twigs are the lifelines supplying the leaves with sap (see fig. 1.3).

1.1.2 The food process (fig. 1.4)

The leaves play the vital role of producing the tree's food. By absorbing daylight energy via the green pigment (chlorophyll) in the leaf, they convert a mixture of carbon dioxide taken from

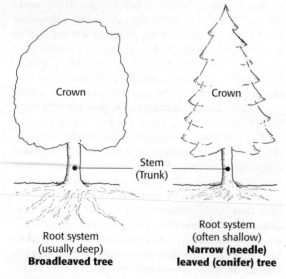

Crown

Crown

Stem
(Trunk)

Root system
(usually deep)
Broadleaved tree

Root system
(often shallow)
**Narrow (needle)
leaved (conifer) tree**

| **Fig 1.2** | General tree shape with their main parts |

both the air and sap from the roots into the necessary amounts of sugars and starches (referred to as food) while, at the same time releasing oxygen into the atmosphere as a waste product. This process is known as *photosynthesis*. However, during the hours of darkness, this action to some extent is reversed – the leaves take in oxygen and give off carbon dioxide, a process known as *respiration* (breathing).

For the whole process to function, there must be some form of built-in system of circulation which allows sap to rise from the ground to the leaves and then to descend as food to be distributed throughout the whole tree. It would seem that this action is due either to suction, induced by *transpiration* (leaves giving off moisture by evaporation), and/or to capillarity (a natural tendency for a liquid to rise within the confines of the cells – see Table 1.12) within the cell structure of the wood.

1.1.3 Structural elements of a tree

We will start from the outer circumference of the tree and work towards its centre. The following features are illustrated in fig. 1.5.

a **Bark** – the outer sheath of the tree. It functions as:
- a moisture barrier,
- a thermal insulator, against extremes in temperature – both hot or cold.
- an armour plate against extremes of temperature, attack by insects, fungi, and animals.

The bark of a well-established tree can usually withstand minor damage, although excessive ill treatment to this region could prove fatal.

b **Inner bark** (bast or phloem) – Conducts food throughout the whole of the tree, from the leaves to the roots.

c **Cambium** (fig. 1.6) – A thin layer or sleeve of cells located between the sapwood and the bast (phloem). These cells are responsible for the tree's growth. As they are formed, they become subdivided in such a way that new cells are added to both sapwood and phloem, thus increasing the girth of the tree.

d **Growth ring** – (Sometimes referred to as an annual ring) – wood cells that have formed around the circumference of the tree during its growing season. The climate and time of

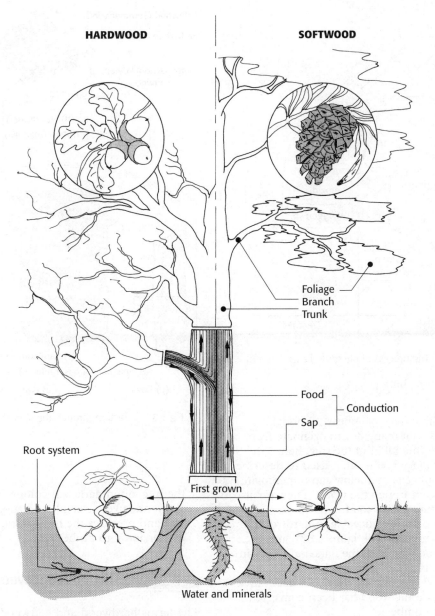

HARDWOOD

SOFTWOOD

Foliage
Branch
Trunk

Food ⎤
        ⎦ Conduction
Sap ⎤

Root system

First grown

Water and minerals

Fig 1.3 Growth of a tree (hardwood & softwood)

year dictate the growth pattern. Each ring is often seen as two distinct bands, known as *earlywood* (springwood) and *latewood* (summerwood). Latewood is usually more dense than earlywood and can be recognised by its darker appearance.

Growth rings are important because they enable the woodworker to decide on the suitability of the wood as a whole – either as timber for joinery (appearance & stability) or, its structural properties (strength & stability) as carcaseing timber.

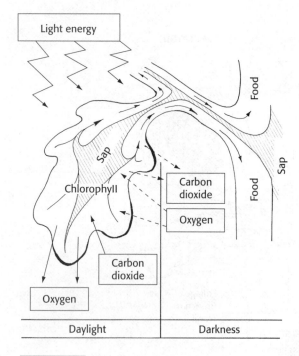

Fig 1.4   The process of photosynthesis

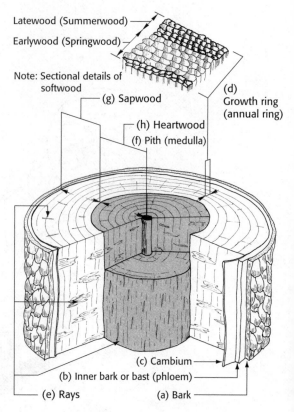

Fig 1.5   Section through the stem/trunk

e **Rays** – these may all appear (although falsely – as not many do) to originate from the centre (medulla) of the tree, hence the term *medullary rays* is often used to describe this strip of cells that allow sap to percolate transversely through the wood. They are also used to store excess food.

Rays are more noticeable in hardwood than in softwood (see figure 1.83), and even then can be seen with the naked eye only in such woods as oak and beech. fig. 1.21 shows how rays may be used as a decorative feature once the wood has been converted (sawn into timber).

f **Pith (medulla)** – the core or centre of the tree, formed from the tree's earliest growth as a sapling. Wood immediately surrounding the pith is called *juvenile wood*, which is not suitable as timber.

g **Sapwood** – the outer active part of the tree which, as its name implies, receives and conducts sap from the roots to the leaves. As this part of the tree matures, it gradually becomes heartwood.

h **Heartwood** – the natural non-active part of

the tree, often darker in colour than sapwood, gives strength and support to the tree and provides the most durable wood for conversion into timber.

1.2   Hardwood & softwood trees

The terms hardwood and softwood can be very confusing, as not all commercially classified hardwoods are physically hard, or softwoods soft. For example, the obeche tree is classed as a hardwood tree, yet it offers little resistance to a saw or chisel etc. The yew tree, on the other hand, is much harder to work yet it is classified as a softwood. To add to this confusion we could be led to believe that hardwood trees are decid-uous (shed their leaves at the end of their grow-ing season) and softwood trees are evergreen (retain their leaves for more than one year),

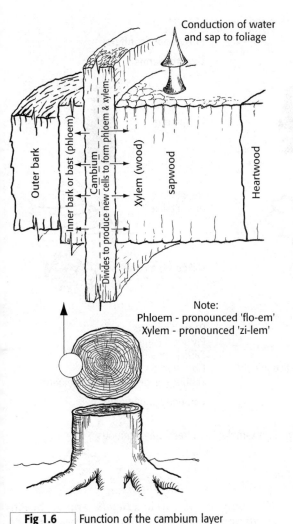

Conduction of water and sap to foliage

Note:
Phloem - pronounced 'flo-em'
Xylem - pronounced 'zi-lem'

| **Fig 1.6** | Function of the cambium layer |

which is true of most species within these groups, but not all!

Table 1.1 identifies certain characteristics found in hardwood and softwood trees; however, it should be used only as a general guide.

Hardwood and softwood in fact, refer to botanical differences in cell composition and structure (Cell types and their formation are dealt with in section 1.10).

1.2.1 Tree & timber names

Common names are often given to trees, (and other plants) so as to include a group of similar yet botanically different species. It is these com-

mon English names which are predominantly used in our timber industry. The true name or Latin botanical name of the tree must be used where formal identification is required – for example:

	Species (true name or Latin botanical <u>name</u>)	
	Genus	
Common name	(generic name	Specific name
English name	or 'surname')	(or 'forename')
Scots pine	*Pinus*	*sylvestris*

As a general guide, it could therefore be said that plants have both a surname and a forename, and, to take it a step further, belong to family groups of hardwood and softwood.

Commercial names for timber often cover more than one species. In these cases, the botanical grouping is indicated by '**spp**', telling us that similar species may be harvested and sold under one genus (singular of genera).

Table 1.2 should give you a better idea how family groups are formed. The first division is into hardwoods and softwoods, next, into their family group, this is followed by their 'Genera' group, then finally their species.

1.3   Forest distribution (source and supply of timber)

The forests of the world that supply the wood for timber, veneers, wood pulp, and chippings for particle board, are usually situated in areas which are typical for a particular group of tree species. For example, as can be seen from figure 1.7, the coniferous forests supplying the bulk of the world's softwoods are mainly found in the cooler regions of northern Europe, also Canada and Asia – stretching to the edge of the Arctic Circle. Hardwoods, however, come either from a temperate climate (neither very hot nor very cold) – where they are mixed with faster-growing softwoods – or from subtropical and tropical regions, where a vast variety of hardwoods grow.

There is increasing concern about the environmental issues concerning forest management, particularly with regard to over extraction of certain wood species – many of which are now protected. This concern has led to some suppliers certifying that their timber is from a

Table 1.1 Guide to recognising hardwood and softwood trees and their use

Botanical grouping	Hardwoods Angiosperms	Softwoods (conifers) Gymnosperms
Leaf group	Deciduous* and evergreen	Evergreen†
Leaf shape	Broadleaf	Needle leaf or scale like
Seed	Encased	Naked via a cone
General usage	Paper and card Plywood (veneers and core) Particle board Timber – heavy structural, decorative joinery	Paper and card Plywood (veneer and core) Particle board Fibre board Timber – general structural joinery
Trade use	Purpose made joinery Shopfitting	Carpentry and joinery

Note: *Within temperate regions around the world; †Not always, for example: larch trees are deciduous

non-protected or sustainable source. Suppliers are now being urged to specify that any timber or wood product used, should be able to present a copy of their Environmental Policy with regard to the products they are to provide.

1.3.1 Temperate hardwoods

These hardwood trees are found where the climate is of a temperate nature. The temperate regions stretch north and south from the tropical areas of the world, into the USSR, Europe, China and North America in the Northern Hemisphere, and Australia, New Zealand and South America in the Southern Hemisphere.

The United Kingdom is host to many of these trees, but not in sufficient quantities to meet all its needs. It must therefore rely on imports from countries which can provide such species as oak (*Quercus spp.*), sycamore (*Acer* spp.), ash (*Fraxinus. spp.*), birch (*Betula* spp.), beech (*Fagus* spp.), and elm (*Ulmus* spp.) – *which* is now an endangered species due to Dutch elm disease.

1.3.2 Tropical and subtropical hardwoods

Most tropical hardwoods come from the rain forests of South America, Africa, and South East Asia. Listed below are some hardwoods which are commonly used:

African mahogany (*Khaya* spp.)	– West Africa
†Afrormosia (*Pericopsis elata*)	– West Africa
Agba (*Gossweiterodendron balsamiferum*)	– West Africa

Table 1.2 The family tree

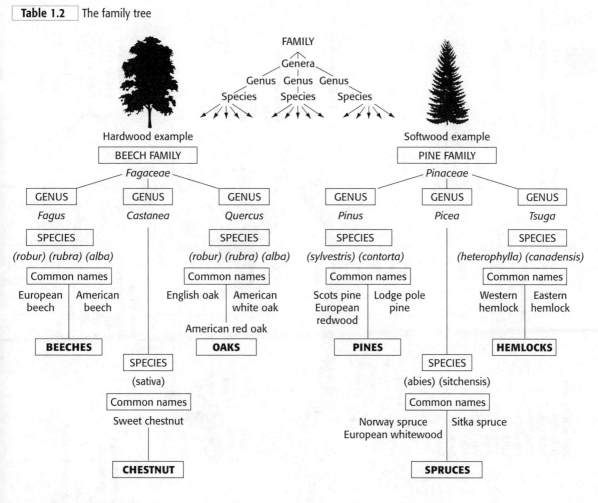

‡American mahogany
(Swietenia macrophylla)   – Central & South
                                America
Gaboon *(Aucoumea*   – West Africa
klaineana)
Iroko *(Chlorophora*   – West Africa
excelsa)
Keruing *(Dipterocarpus* spp.) – South East Asia
Meranti *(Shorea* spp.)   – South East Asia
Sapele   – West Africa
(Entandrophragma
cylindricum)
Teak *(Tectona grandis)*   – Burma, Thailand
Utile *(Entandophragma utile)* – West Africa

N.B. Many of these species may have originated
from tropical rain forest regions which the timber
industry is trying to control as conservation areas:
† = Trading restrictions

‡ = *Protected species*
Also, see table 1.15

1.3.3 Hardwood use

Hardwoods may be placed in one or more of the
following purpose groups:

Purpose group	Use
a Decorative	natural beauty – colour and/or figured grain
b General-purpose	joinery and light structural
c Heavy structural	withstanding heavy loads

1.3.4 Softwoods

Most timber used in the UK for carpentry and
joinery purposes is softwood imported from

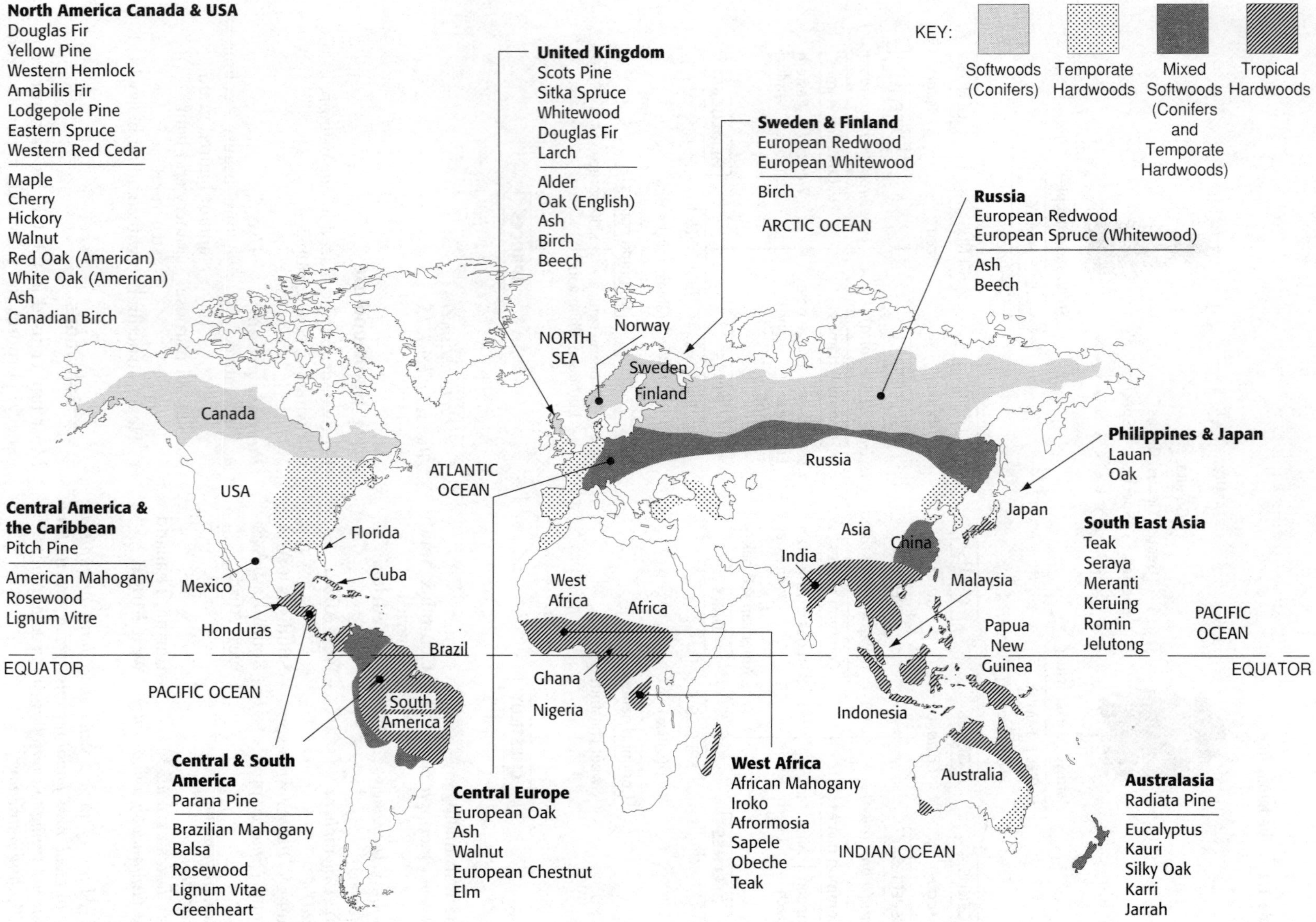

Fig 1.7 Forest regions of the world

North America Canada & USA
Douglas Fir
Yellow Pine
Western Hemlock
Amabilis Fir
Lodgepole Pine
Eastern Spruce
Western Red Cedar

Maple
Cherry
Hickory
Walnut
Red Oak (American)
White Oak (American)
Ash
Canadian Birch

Central America & the Caribbean
Pitch Pine

American Mahogany
Rosewood
Lignum Vitre

Central & South America
Parana Pine

Brazilian Mahogany
Balsa
Rosewood
Lignum Vitae
Greenheart

Central Europe
European Oak
Ash
Walnut
European Chestnut
Elm

United Kingdom
Scots Pine
Sitka Spruce
Whitewood
Douglas Fir
Larch

Alder
Oak (English)
Ash
Birch
Beech

Sweden & Finland
European Redwood
European Whitewood

Birch

Russia
European Redwood
European Spruce (Whitewood)

Ash
Beech

Philippines & Japan
Lauan
Oak

South East Asia
Teak
Seraya
Meranti
Keruing
Romin
Jelutong

West Africa
African Mahogany
Iroko
Afrormosia
Sapele
Obeche
Teak

Australasia
Radiata Pine

Eucalyptus
Kauri
Silky Oak
Karri
Jarrah

KEY:

Softwoods (Conifers)

Temperate Hardwoods

Mixed Softwoods (Conifers and Temperate Hardwoods)

Tropical Hardwoods

ARCTIC OCEAN

ATLANTIC OCEAN

PACIFIC OCEAN

INDIAN OCEAN

EQUATOR

NORTH SEA

Canada
USA
Mexico
Honduras
Florida
Cuba
Brazil
South America
Norway
Sweden
Finland
Russia
Asia
China
Japan
India
Malaysia
Papua New Guinea
Indonesia
Australia
Africa
West Africa
Ghana
Nigeria

Sweden, Finland, and the USSR. The most important of these softwoods are European redwood *(Pinus sylvestris)*, which includes Baltic redwood, and Scots pine, a native of the British Isles. As timber, these softwoods are collectively called simply 'redwood'. Redwood is closely followed in popularity by European Whitewood, a group which includes Baltic Whitewood and Norway spruce *(Picia abies)* – recognised in the UK as the tree most commonly used at Christmas as the Christmas tree. Commercially, these, and sometimes silver firs are simply referred to as 'whitewood'.

Larger growing softwoods are found in the pacific coast region of the USA and Canada. These include such species as Douglas fir *(Pseudotsuga menziesii)* – known also as Columbian or Oregon pine, although technically not a pine. Western hemlock *(Tsuga heterophylla)*, and Western red cedar *(Thuj'a plicata)*. Western red cedar is known for its durability and its resistance to attack by fungi. Brazil is the home of Parana pine *(Araucaria angustifolia)*, which produces long lengths of virtually knot-free timber, which is however, only suitable for interior joinery purposes.

1.3.5 Forms of supply

Softwood is usually exported from its country of origin as sawn timber in packages, or in bundles. It has usually been pre-dried to about 20% m.c. (Moisture content – see section 1.7). Packaged timber is to a specified quality and size, bound or bonded with straps of steel or plastics for easy handling, and wrapped in paper or plastics sheets.

Hardwood, however, may be supplied as sawn boards or as logs to be converted (sawn) later by the timber importer to suit the customer's requirements.

1.4   Conversion into timber

Felling (the act of cutting down a living tree) is carried out when trees are of a commercially suitable size, having reached maturity, or for thinning-out purposes. Once the tree has been felled, its branches will be removed, leaving the trunk (stem) in the form of a log. The division of this log into timber sections is called *conversion*.

What, then, is the difference between *wood* and *timber*? The word wood is often used very loosely to describe timber, when it should be used to describe either a collection of growing trees or the substance that trees are made of, i.e. the moisture-conducting cells and tissues etc. Timber is wood in the form of squared boards or planks etc.

Initial conversion may be carried out in the forest whilst the log is in its green (freshly felled) state by using heavy, yet portable machines, such as circular saws or vertical and horizontal band-mills (see fig. 1.14). This leads to a reduction in transport cost, as squared sectional timber can be transported more economically than logs.

Alternatively, the logs may be transported by road, rail, or water to a permanently sited sawmill. Were they are kept wet, either within a log pond, or with water sprinklers.

1.4.1 Sawing machines

The type of sawing equipment used in a sawmill will depend on the size and kind of logs it handles. For example:

a *Circular saw* (fig. 1.8) – small- to medium diameter hardwoods and softwoods. Figure 1.8 shows a rolling table log saw. The tables are available in lengths from 3.05 m to 15.24 m, and the diameter of saw could be as large as 1.829 m.

b *Vertical frame saw or gang saw* (fig. 1.9) – small to medium-diameter softwoods. The log is fed and held in position by fluted rollers while being cut with a series of reciprocating upward-and-downward

| **Fig 1.8** | Circular saw |

Fig 1.9 Vertical frame saw or gang saw

Fig 1.11 Vertical band-mill

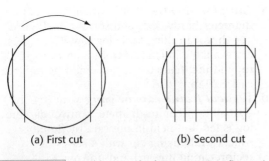

(a) First cut   (b) Second cut

Fig 1.10 Possible cuts of a frame saw (see fig 1.13)

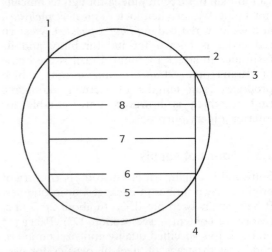

Fig 1.12 Band-mill cuts

moving). saw blades. The number and position of these blades will vary according to the size and shape of each timber section. Figure 1.10(a) illustrates the possible result after having passed the log through this machine once, whereas figure 1.10(b) shows what could be achieved after making a further pass.

c *Vertical band-mill* (fig. 1.11) – all sizes of both hardwood and softwood. Logs are fed by a mechanised carriage to a saw blade in the form of an endless band, which revolves around two large wheels (pulleys), one of which is motorised. Figure 1.12 shows an example of how these cuts can be taken.

d *Double vertical band-saw* (fig. 1.13) – small to medium logs. It has the advantage of making two cuts in one pass.

e *Horizontal band-saw* (fig. 1.14) – all sizes of hardwood and softwood. The machine illustrated is suitable for work at the forest site, or in a sawmill. Conversion is achieved by passing the whole mobile saw unit (which travels on rails) over a stationary log, taking a slice off at each forward pass.

The larger mills may employ a semi-computerised system of controls to their machinery, thus helping to cut down some human error and providing greater safety to the whole operation.

Bansaw blades

Fig 1.13   Double vertical band-saw

Fig 1.14   'Forester-150' horizontal band-mill – through and through sawing

Fig 1.15   Resawing timber

The final control and decisions, however, are usually left to the expertise of the sawyer (machine operator).

Timber which requires further reduction in size is cut on a resaw machine. Figure 1.15 shows a resawing operations being carried out,

one a single unit, and the other using two machines in tandem to speed up the operation.

Importers of timber in the United Kingdom may specialise in either hardwoods or softwoods, or both. Their sawmills will be geared to meet their particular needs, by re-sawing to customers' requirements. Hardwood specialists usually have their own timber drying facilities.

1.4.2 **Method of conversion**

The way in which the log is cut (subdivided) will depend on the following factors:

- type of sawing machine,
- log size (diameter or girth)
- type of wood,
- condition of the wood – structural defects etc.(see section 1.6),
- proportion of heartwood to sapwood,
- future use – structural, decorative, or both.

Broadly speaking, the measures taken to meet the customer's requirements will (with the exception of the larger mills) be the responsibility of the experienced sawyer (as mentioned earlier), whose decision will determine the method of conversion, for example:

a **Through-and-through-sawn** (fig. 1.16) In this method of conversion, parallel cuts are made down the length of the log, producing a number of 'quarter' and 'tangential' sawn boards (figs 1.17 and 1.19). The first and last cuts leave a portion of wood called a 'slab'. This method of conversion is probably the simplest and least expensive.
NB. *Cuts may be made vertically or horizontally depending on the type of machine.*

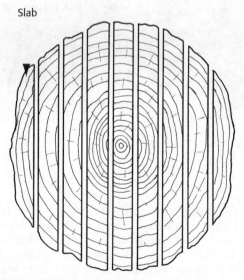

Slab

Fig 1.16   Through and through sawn (producing plain and quarter sawn timber)

b *Tangential-sawn* (Plain sawn) – figure 1.17 shows that by starting with a squared log, tangential-sawn boards are produced by working round the log, by turning it to produce boards, all of which (except the centre) have their growth rings across the boards' width. Figure 1.18 shows alternative methods leaving a central 'boxed heart'
Although tangential-sawn sections are

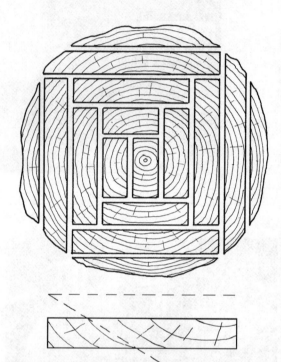

Plain sawn timber - growth rings meet the face of the board at an angle less than 45°

Fig 1.17   Tangentially sawn (producing 'plain sawn' or 'flat sawn' timber – except for heartboards)

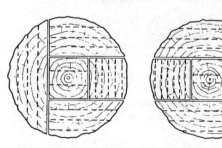

Fig 1.18   Dividing the log to produce plain sawn timber and a boxed heart

subject to cupping (becoming hollow across the width) when they dry, they are the most suitable sections for softwood beams, i.e. floor joists, roof rafters, etc., which rely on the position of the growth ring to give greater strength to the beam's depth.

c **Quarter (radial) or rift-sawn** (fig. 1.19) –

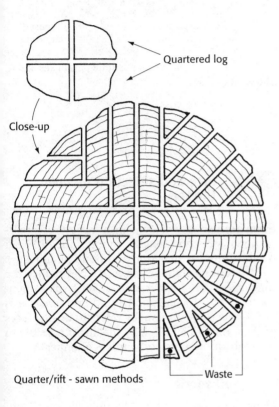

Close-up

Quartered log

Quarter/rift - sawn methods

Waste

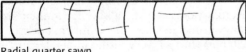

Radial quarter sawn

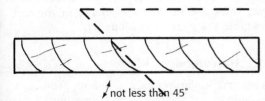

not less than 45°

Acceptable quarter sawn

Fig 1.19   Quarter (radial) or rift sawn timber

this method of conversion can be wasteful and expensive, although it is necessary where a large number of radial or near radial-sawn boards are required. Certain hardwoods cut in this fashion, produce beautiful figured boards (fig. 1.21), for example, figured oak, as a result of the rays being exposed (fig. 1.5). Quarter-sawn boards retain their shape better than tangential-sawn boards and tend to shrink less, making them well suited to good-class joinery work and quality flooring.

The resulting timber, with the exception of that which surrounds the 'heart wood' shown in table 1.3 will either be:

● Tangentially (plain) sawn.
● Quarter sawn or Rift sawn.

Table 1.3   Comparison between 'plain' and 'quarter' sawn timber

Advantages	Disadvantages
Plain sawn	
Economical conversion	Tends to 'cup' (distort) on drying due to shrinkage – ('cupping' is its natural pattern of shrinkage)
Ideal section for softwood beams	
Can produce a decorative pattern (flower or flame figure) on the tangential face of the timber with distinct growth rings – see Figures 1.21 and 1.86c	
Quarter sawn	
Retains it shape better during drying	Expensive form of conversion
Shrinkage across its width half of that of plain sawn timber	Conversion methods can be wasteful
Ideal selection for flooring with good surface wearing properties	
Produces a decorative radial face (e.g. Silver Figure) on hardwoods with broad ray tissue, see figures 1.21 and 1.86 b and c	

1.4.3 Conversion geometry (Fig. 1.20)

Knowing that a log's cross-section is generally just about circular, the above-mentioned saw cuts and sections could be related to a circle and its geometry. For example, timber sawn near to a 'radius' line will be *radial-sawn*. quartered logs (divided by cutting into four quarters) or *quadrants*. Similarly, any cut made as a tangent to a growth ring would be called *tangentially sawn*. The *'chords'* are straight lines, which start and finish at the circumference; therefore a series of chords can be related to a log that has been sawn 'through-and-through', 'plain sawn' or 'flat sawn'. It should be noted that the chord line is also used when cuts are made tangential to a growth ring, and when the log is cut in half.

1.4.4 Decorative boards

Figure 1.21 gives two examples of how wood can be cut to produce timber with an attractive face.

Quarter sawn hardwoods with broad rays can produce nicely figured boards. For example, quarter sawn European Oak is well known for its 'Silver figure' when sawn in this way. Tangentially sawn softwoods with distinct growth rings can produce a flame like pattern on their surface – known as 'Flame figuring'.

Further examples can be see in figure 1.86.

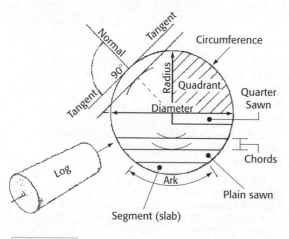

Fig 1.20   Conversion geometry

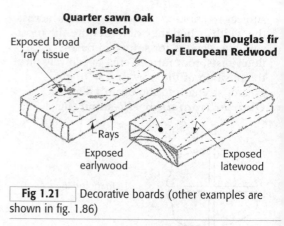

Fig 1.21   Decorative boards (other examples are shown in fig. 1.86)

1.5   Size and selection of sawn timber)

Sawn timber is available in a variety of cross-sectional sizes and lengths to meet the different needs of the construction and building industry.

By adopting standard sizes, we can reduce the time spent on further conversion, subsequent wastage, and the inevitable build-up of *short ends* or off-cuts (off-cuts usually refers to waste pieces of sheet materials), thereby making it possible to plan jobs more efficiently and economically.

1.5.1 Softwoods

Depending on whether the suppliers are from North America or Europe, stated cross- section sizes can vary. Canadian mills, unlike European mills, may not make any allowance in their sizes for any shrinkage when their timber is dried.

Timber shrinks very little in its length, so allowance provisions are not necessary.

Table 1.4(a) shows the cross sectional sizes of sawn softwood normally available in the U.K. and table 1.4(b) their cut lengths.

1.5.2 Hardwoods

As shown in figure 1.22 different profiles are available to suit the end user. Dimentioned sawn stock sizes as shown in table 1.5 may be available, but this will depend on species and local availability.

Table 1.4   Sawn sizes of softwood timber

(a) Customary target sizes of sawn softwood

Thickness (mm)	Width (mm)											
	75	100	115	125	138	150	175	200	225	250	275	300
16	X	X		X		X						
19	X	X		X		X						
22	X	X		X		X						
25	X	X		X		X	X	X	X	X	X	X
32		X	X	X	X	X	X	X	X	X	X	X
38	X	X	X	X	X	X	X	X	X	X	X	X
47	X	X		X		X	X	X	X	X		X
50	X	X		X		X	X	X	X	X		X
63		X		X		X	X	X	X			
75		X		X		X	X	X	X	X	X	X
100		X				X		X	X	X	X	X
150						X		X				X
250										X		
300												X

Note: Certain sizes may not be obtainable in the customary range of species and grades which are generally available. Permitted deviation of cross-sectional sizes at 20% moisture content.
- for thickness and widths ≤100 mm $\left[^{+3}_{-1}\right]$ mm;
- for thickness and widths >100 mm $\left[^{+4}_{-2}\right]$ mm.
Target size of 20% moisture content.

(b) Customary lengths of sawn softwood

1.80	2.10	3.00	4.20	5.10	6.00	7.20
	2.40	3.30	4.50	5.40	6.30	
	2.70	3.60	4.80	5.70	6.60	
		3.90			6.90	

Note: Lengths of 5.70 m and over may not be readily available without finger jointing

(c)

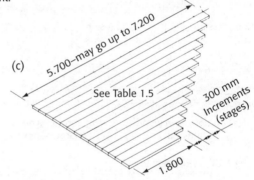

5.700–may go up to 7.200
See Table 1.5
300 mm Increments (stages)
1.800

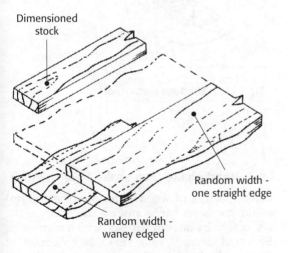

Dimensioned stock

Random width - one straight edge

Random width - waney edged

Fig 1.22   Profiles of hardwood sections

1.6   Structural defects (natural defects)

Figures 1.23 to 1.27 show defects that may be evident before, and/or during conversion. Most of these defects have little, if any, detrimental effect on the tree, but they can degrade the timber cut from it, i.e. lower its market value.

1.6.1 Reaction wood (fig 1.23)

This defect is the result of any tree which has had to grow with a natural leaning posture, this may be as a result of having to resist strong prevailing winds, or having to established itself on sloping ground. These trees resist any pressure existed upon them by attempting to grow vertically with

Table 1.5   Basic guide sizes of sawn Hardwood

Thickness (mm)	Width (mm)										
	50	63	75	100	125	150	175	200	225	250	300
19			×	×	×	×	×				
25	×	×	×	×	×	×	×	×	×	×	×
32		×	×	×	×	×	×	×	×	×	×
38			×	×	×	×	×	×	×	×	
50				×	×	×	×	×	×	×	×
63						×	×	×	×	×	×
75						×	×	×	×	×	×
100						×	×	×	×	×	×

Note: Designers and users should check the availability of specified sizes in any particular species

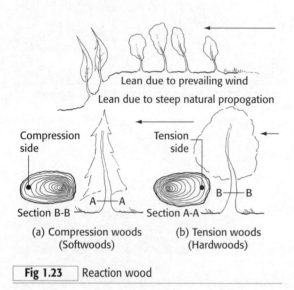

(a) Compression woods (Softwoods)   (b) Tension woods (Hardwoods)

Fig 1.23   Reaction wood

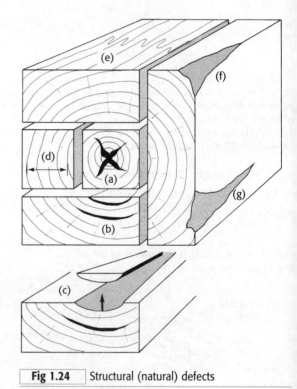

Fig 1.24   Structural (natural) defects

added supportive wood growth to their stem. This extra growth will be formed in such a way that the stem will take on an eccentric appearance around the stem, which, with softwood is on the side of the tree that is being subjected to compressive forces – this wood is known as a compression wood. Hardwoods on the other hand, produce extra wood on the side likely to be stretched, since this is the side in tension. This wood is known as **tension wood**. In both these cases the wood is unsuitable as timber since it would be unstable went dried and particularly hazardous when processed. Collectively, both compression and tension wood are known as **reaction wood**.

1.6.2 Heart shake (Star shake – Fig. 1.24(a))

Shake (parting of wood fibres along the grain) within the heart (area around the pith) of the tree caused by uneven stresses, which increase as the wood dries. A star shake is collection of shakes radiating from the heart.

1.6.3 **Ring shake** (Cup shake – Fig. 1.24(b))

A shake which follows the path of a growth ring. Figure 1.24 (c) shows the effect it can have on a length of timber.

1.6.4 **Natural compression failure** (upset – fig. 1.24(e))

Fracturing of the fibres; thought to be caused by sudden shock at the time of felling or by the tree becoming over-stressed (during growth) – possibly due to strong winds etc.

NB. Other names for this defect include 'thunder shake' or 'lightning shake'.

1.6.5 **Rate of growth** fig. 1.24(d))

The number of growth rings per 25 min, can with softwoods determine the strength of the timber.

1.6.6 **Wane** (Waney-edge – fig. 1.24(f))

The edge of a piece of timber that has retained part of the tree's rounded surface, possibly including some bark.

1.6.7 **Encased bark** (fig. 1.24(g))

Bark may appear inset into the face or the edge of a piece of timber.

1.6.8 **Sloping grain** (fig. 1.25)

The grain (direction of the wood fibres), slopes sharply in a way that can make load-bearing timbers unsafe, e.g. beams and joists.

Figure 1.25a shows possible source. Figure 1.25b a method of testing for sloping grain. Figure 1.25c how sloping grain could be responsible for pre-mature fracturing of a beam.

1.6.9 **Knots**

As shown in figure 1.26 where the tree's branches join the stem they become an integral part of it. The lower branches are often trimmed off by forest management during early growth, this encourages 'clear wood' to grow over the knot as the tree develops.

Figure 1.27 shows how knots may appear in the sawn timber. The size, type, location, and number of knots, are controlling factors when the timber is graded for use.

Some of the terms used to describe knots are:

● *Dead knots* – If a branch is severely damaged, that part adjoining the stem will die and may eventually become enclosed as the tree develops – not being revealed until

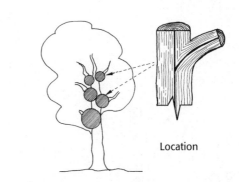

Location

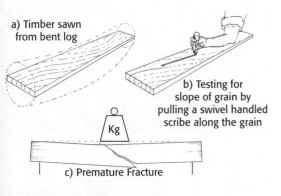

a) Timber sawn from bent log

b) Testing for slope of grain by pulling a swivel handled scribe along the grain

Kg

c) Premature Fracture

Fig 1.25 | Sloping grain

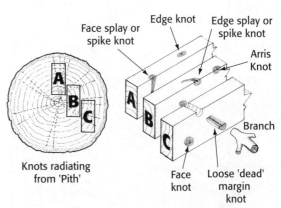

Face splay or spike knot

Edge knot

Edge splay or spike knot

Arris Knot

Branch

Knots radiating from 'Pith'

Face knot

Loose 'dead' margin knot

Fig 1.26 | Knots in relation to branches and stem

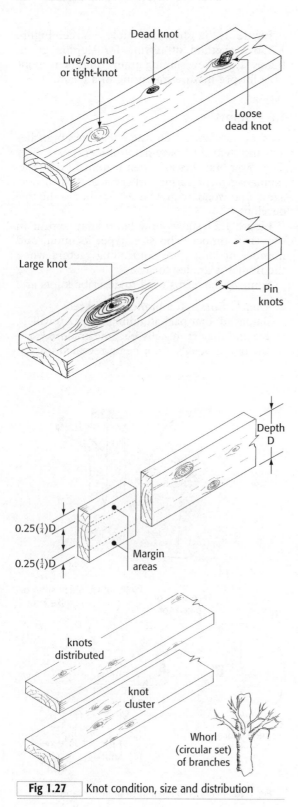

Fig 1.27 Knot condition, size and distribution

conversion into timber. *Note: these knots are often loose, making them a potential hazard whenever machining operations are carried out.*

- **Knot size** – Larger the knot greater the strength reduction of the timber.
- **Knot location** – Knots nearer the edges (margins) of the beam are generally going to reduce the strength properties of timber, rather than those nearer the centre.
- **Number of Knots** – Generally the greater the distance between the knots the better.
- **Knot types** – Knots appearing on the surfaces of timber take many forms the names reflect their position, for example:
  - Face knots
  - Margin knots
  - Edge knots
  - Arris knots
  - Splay or spike knots

1.6.10 Resin (Pitch) pocket

An a opening, following the saucer shape of a growth ring containing an accumulation of resin. Apparent in many softwoods, mainly in spruces – it may appear as a resinous streak on the surface of timber. In warm weather sticky resin may run down vertical members. When the resin dries it takes on a resinous granular form which can be scraped away.

1.7   Drying timber

Timber derived from freshly felled wood is said to be *green,* meaning that the cell cavities *contain free water* and the wall fibres are saturated with bound water (fig. 1.33), making the wood heavy, structurally weak, susceptible to attack by insects and/or fungi, also unworkable. Timber in this condition is therefore always unsuitable for use. The amount of moisture the wood contains as a percentage of the oven-dry weight, is known as the *moisture content* (m. c.), and the process of reducing the m.c is termed *drying.*

The main object of drying timber is therefore to:

- reduce its weight,
- increase its strength properties,
- increase its resistance to fungal and attack by some insects,

- provide stability with regards to moisture movement,
- increase workability for machine and hand tools
- enable wood preservative treatments to be applied, (with the exception of those applied by diffusion – section 3.4.2)
- enable fire retardant treatments to be applied
- enable surface finishes to be applied
- enable adhesives to be applied,
- reduce the corrosive properties of some woods,
- reduce heat conductivity thereby increase thermal insulation properties,

and produce timber with a level of moisture content acceptable for its end use. Examples are given in figure 1.28 and table 1.6.

The drying process, (sometimes called 'seasoning'), is usually carried out by one of three methods:

a Air-drying (natural drying),
b Kiln-drying (artificial drying),
c Air-drying followed by kiln-drying.

All three methods aim at producing timber that will remain stable in both size and shape – the overriding factor being the final moisture content, which ultimately controls the use of the timber.

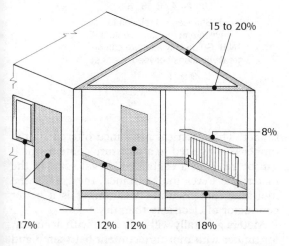

15 to 20%

8%

17%   12% 12%   18%

N.B. Wood with a 20% + M.C. is liable to attack by fungi

Fig 1.28   Moisture content of wood products in various situations

Although outside the scope of this book other drying methods include:

- forced-air drying,
- climate chambers,
- dehumidifiers,
- vacuum drying,

Because the object of drying timber is to remove water from the cells (fig. 1.33), moisture content is considered first.

1.7.1 Moisture content (m.c)

As already expressed, the moisture content of wood is the measured amount of moisture within a sample of wood expressed as a percentage of its dry weight. If the weight of water present exceeds that of dry wood, then moisture contents of over 100% will be obtained.

There are several methods of determining moisture content values, but we will only be considering the following two methods:

- the traditional oven-drying method, and
- using modern electrical moisture meters and probes,

a **Oven-drying method** (fig. 1.29) – a small sample cut from the timber which is to be dried (see fig. 1.43) is weighed to determine its 'initial' or 'wet' weight. It is then put into an oven with a temperature of 103°C ± 2°C until no further weight loss is recorded, its weight at this stage being known as its 'final' or 'dry' weight.

Once the 'wet' and 'dry' weights of the sample are known, its original moisture content can be determined by using the following formulae:

Moisture content % =

$$\frac{\text{Initial (wet weight } (A)) - \text{final (dry weight } (B))}{\text{Final or dry weight } (B)} \times 100$$

Or

$$\text{MC \%} \frac{\text{Initial (wet weight } (A))}{\text{Final or dry weight } (B)} - 1 \times 100$$

For example, if a sample has a wet weight of 25.24 g (A) and a dry weight of 19.12 g (B), then:

$$\text{MC \%} = \frac{A - B}{B} \times 100$$

$$\frac{25.24 \text{ g} - 19.12 \text{ g}}{19.12 \text{ g}} \times 100 = 32\%$$

Table 1.6 Moisture content of timber in relation to its end use

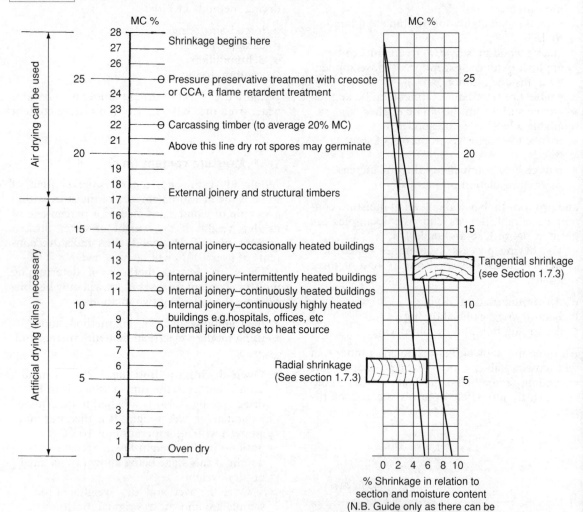

% Shrinkage in relation to
section and moisture content
(N.B. Guide only as there can be
great variations between species)

Or

$$MC\ \% = \left[\left(\frac{A}{B}\right) - 1\right] \times 100$$

$$MC\ \% = \left[\left(\frac{25.24\ g}{19.12\ g}\right) - 1\right] \times 100 = 32\%$$

b Electrical moisture meters (fig. 1.30)

These are battery-operated instruments, which usually work by relating the electrical resistance of timber to the moisture it contains.

Moisture content is measured by pushing or driving (hammer-type) two electrodes into the timber. The electrical resistance offered by the timber is converted to a moisture content which can be read off a calibrated digital scale of the meter, the lower the resistance, the greater the moisture content, since wet timber is a better conductor of electricity than dry.

Meters generally will only cope with accuracy, for timber with moisture content between 6 and 28%. Above this point (the fibre-saturation point) there will be little or no change in electrical resistance. With the exception of the small hand held models (fig 1.31) useful for making

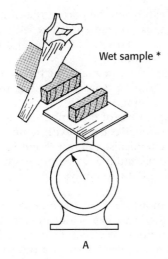

Wet sample *

A

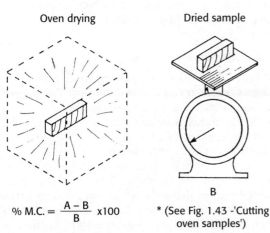

Oven drying

Dried sample

B

$$\% \text{ M.C.} = \frac{A - B}{B} \times 100$$

* (See Fig. 1.43 -'Cutting oven samples')

Fig 1.29 Method of determining moisture content by oven drying a small sample of timber (also see fig 1.43)

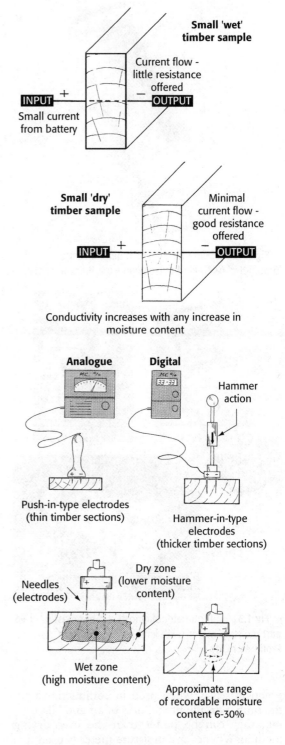

Small 'wet' timber sample

Current flow - little resistance offered

INPUT + − OUTPUT

Small current from battery

Small 'dry' timber sample

Minimal current flow - good resistance offered

INPUT + − OUTPUT

Conductivity increases with any increase in moisture content

Analogue   Digital

Hammer action

Push-in-type electrodes (thin timber sections)

Hammer-in-type electrodes (thicker timber sections)

Needles (electrodes)

Dry zone (lower moisture content)

Wet zone (high moisture content)

Approximate range of recordable moisture content 6-30%

Fig 1.30 Battery operated moisture meter

spot checks on site. Moisture meters are in two parts (fig 1.32):

- The meter itself with both a numerical scale and pointer, or digital readout – provision will be made for adjustment to suit different wood species – this part will also have provision for housing the batteries. It may also, like the one shown in figure 1.32 have integral pins to allow surface readings to be taken.
- Spiked electrodes (probes) set into an insulated hand-piece, with provision for attaching it to the meter via a detachable cable.

Moisture meter systems are more than just a useful aid for making spot checks – in fact in the

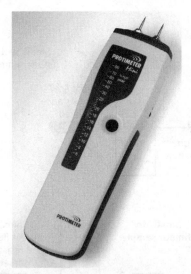

Fig 1.31 Hand held 'mini' moisture metre by 'protimeter' (with kind permission from Protimeter Ltd)

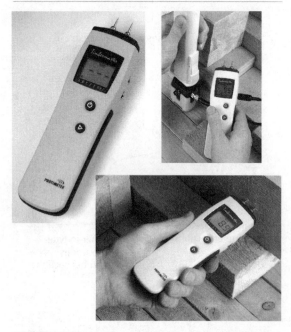

Fig 1.32 Protimeter diagnostic timber master – two part moisture meter (with kind permission from Protimeter Ltd)

practical sense, when used in conjunction with the timber drying procedures of air and kiln drying, they can be better than the oven-drying method. Whenever a moisture meter is used it is important that:

a probes can reach the part of the timber whose moisture content is needed (depends on the sectional size of the timber and the type of instrument);

b allowance must be made for the timber species – timber density can affect the meter's reading;

c the temperature of the timber is known – meter readings can vary with temperature;

d certain chemicals are not present in the timber, for example, wood preservatives or flame-retardant solutions.

Tests on moisture content may be necessary when sorting large batches of timber, or checking the condition of assembled or fixed carpentry and joinery, particularly, if a fungal attack is in evidence or suspected – in which case a moisture meter would be invaluable.

1.7.2 Moisture removal

Before considering the two main drying techniques, let us try to understand how this loss of moisture may effect the resulting timber. Figure 1.33 illustrates how moisture is lost naturally, and the effect it can have on a timber section if moisture is then reintroduced.

We already know that green timber contains a great amount of water. This water is contained within the cell cavities – we call this free water, because it is free to move around from cell to cell. The water contained within the cell walls is fixed (chemically bound to them), and is therefore known as bound water or bound moisture.

As you will know the air we breathe contains varying amounts of moisture: the amount will depend on how much is suspended in the air as vapour at that point in time, which in turn will depend on the surrounding air temperature. As the air temperature increases, so does its capacity to absorb more moisture as vapour, until the air becomes saturated, at which point we are very aware of how humid it has become. It is therefore this relationship between air temperature, and the amount of moisture the air can hold that we call relative humidity.

If the air, surrounding the timber has a vacant capacity for moisture, it will take up any spare moisture from the wood until, eventually, the moisture capacity of the air is in balance, or equilibrium, with that of the timber. When stable

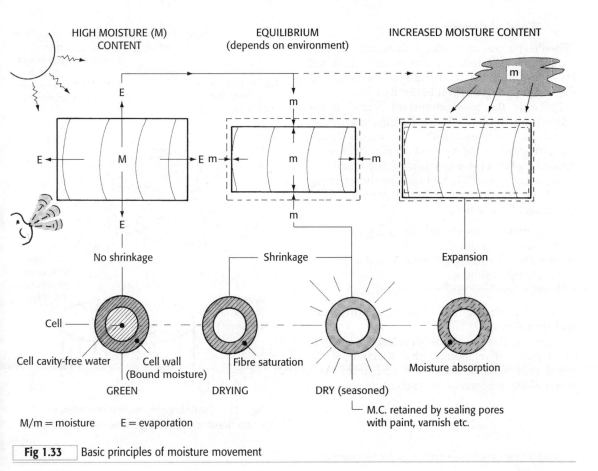

HIGH MOISTURE (M) CONTENT

EQUILIBRIUM (depends on environment)

INCREASED MOISTURE CONTENT

No shrinkage

Shrinkage

Expansion

Cell

Cell cavity-free water

Cell wall (Bound moisture)

Fibre saturation

Moisture absorption

GREEN

DRYING

DRY (seasoned)

└ M.C. retained by sealing pores with paint, varnish etc.

M/m = moisture   E = evaporation

Fig 1.33   Basic principles of moisture movement

conditions are reached we can say an **equilibrium moisture content (EMC)** has been achieved. This process will of course act in reverse, because wood is a **hygroscopic material**, which means that it has the means, provided the conditions (those mentioned above), are suitable, to pick up from and shed moisture to its surrounding environment.

Any free water will leave first, via tiny perforations within the cell walls. As the outer cells of the timber start to dry, they will be replenished by the contents of the inner cells, and so on, until only the cell walls remain saturated. The timber will start to shrink at this important stage of drying, known as the **fibre saturation point (FSP)** when about 25% to 30% m.c. (table 1.6) will be reached.

Beyond fibre saturation point (FSP), drying out bound water can be very lengthy process if left to take place naturally. To speed up the process, artificial drying techniques will need to be employed.

It is worth pointing out at this stage, that it is possible for timber in a changeable environment to remain stable if moisture absorption can be prevented. This may be achieved by one of two methods:

1   Completely sealing all its exposed surfaces,
2   Using a micro-pore sealer that prevents direct entry of water from outside but allows trapped moisture to escape.

All timber must of course be suitably dried before any such treatments are carried out.

1.7.3 Wood shrinkage

Whether natural or artificial means are used to reduce the moisture content of timber, it will inevitably shrink. The amount of shrinkage will depend on the reduction below its FSP.

Probably the most important factor, is the relationship between the differing amounts of shrinkage, compared with the timbers length (longitudinally), and its cross-section (transverse section), whether it is plain saw (tangentially) or quarter saw (radially). And how, as shown in figure 1.34 we can view different proportions of shrinkage, for example:

a tangentially – responsible for the greatest amount of shrinkage
b radially – shrinkage of about half that of tangential shrinkage
c longitudinally – hardly any shrinkage.

We call varying amounts of shrinkage **differential shrinkage**.

Figure 1.35 shows how shrinkage movement takes place in relation to the direction of the wood cells situated across the end grain. As a result of this movement, we can expect some sec-

Radial cell shrinkage reduced to about half of tangential cell movement - this restriction is due in part to lack of movement of ray cells radially

Greatest shrinkage accross the cells tangentially

Ray cells (little or no movement radially)

Axial cells (little or no Movement)

Fig 1.35 | Shrinkage movement in relation to direction of wood cells (exaggerated view of end grain)

a tangentially - responsible for the greatest amount of shrinkage
b radially - shrinkage of about half that of tangential shrinkage
c longitudinally - hardly any shrinkage

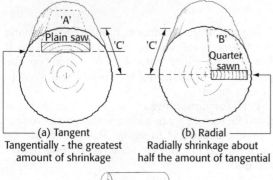

(a) Tangent
Tangentially - the greatest amount of shrinkage

(b) Radial
Radially shrinkage about half the amount of tangential

(c) Length (longitudinal) minimal shrinkage (least amount)

Fig 1.34 | Proportions of wood shrinkage

tions of timber to distort in some way as moisture is removed from the cells to below fibre saturation point. The resulting shapes of distorted timber sections will depend on where the timber was cut out of the log during its conversion into timber. Figure 1.36 should give some idea as to how certain sections of timber may end up after being dry – reference should be made to section 1.7.8 which itemises various drying defects.

1.7.4 Air drying (natural drying)

Often carried out in open-sided sheds, where the timber is exposed to the combined action of circulating air and temperature, which lifts and drives away unwanted moisture by a process of evaporation (similar to the drying of clothes on a washing line). A suitable reduction in m. c. can take many months, depending on:

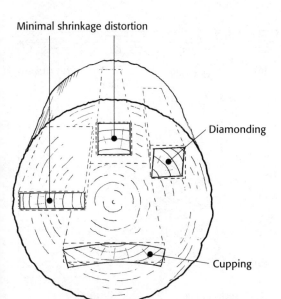

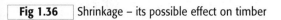

Fig 1.36 | Shrinkage – its possible effect on timber

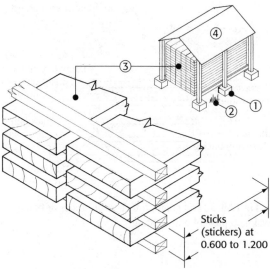

IMPORTANT ELEMENTS:
1. Risen off the ground - no rising damp.
2. Clear of ground vegetation.
3. Free circulation of air.
4. Protection from the weather.

Fig 1.37 | Air drying shelter and stack build-up

a the drying environment and amount of exposure,
b the type of wood (hardwood or softwood),
c the particular species,
d the timber thickness.

The final m. c. obtained can be as low as 16 % to 17 % in summer months and as high as 20 % or more during winter. It would therefore be fair to say that this method of drying timber is very unreliable.

A typical arrangement for air-drying is shown in figure 1.37, where the features numbered are of prime importance if satisfactory results are to be achieved. They are as follows:

1 Timber stacks (piles of sawn timber), must always be raised off the floor, thus avoiding rising damp from the ground. Stacked correctly (fig. 1.39). Concrete, gravel, or ash will provide a suitable site covering.
2 The area surrounding the shed must be kept free from ground vegetation, to avoid conduction of moisture from the ground.
3 Free circulation of air must be maintained throughout the stack – the size and position of 'sticks' will depend on the type, species, and section of timber being dried.

4 The roof covering must be sound, to protect the stacks from adverse weather conditions.

The success of air-drying will depend on the following factors:

a weather protection,
b site conditions,
c stacking as shown in figure 1.39,
d atmospheric conditions.

a Weather protection
Except when drying certain hardwoods which can be dried as an open-piled 'boule' (the log being sawn through-and-through and then reassembled into its original form – see 'Stacking'), a roof is employed to protect the stack from direct rain or snow and extremes in temperature. Its shape is unimportant, but corrugated steel should be avoided in hot climates because of its good heat-conducting properties that would accelerate the drying process. Roof coverings containing iron are liable to rust and should not be used where species of a high tannin content (such as Oak, Sweet chestnut, Afrormosia, Western red cedar, etc.) are being

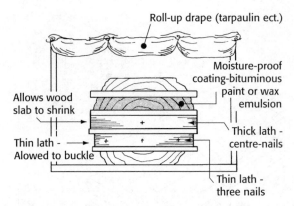

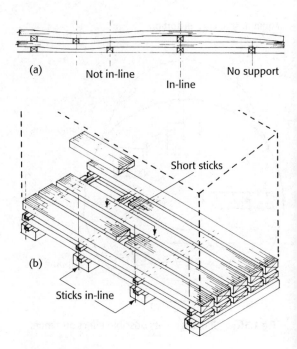

Fig 1.38 End grain protection of timber or boule will help prevent end splitting

Softwood sticks 25mm x 13mm to 25mm x 25mm at intervals of 0.600 to 1.200 centres - depending on board thickness and drying rate

Fig 1.39 Build-up of stack

dried, otherwise iron-staining is possible where roof water has dripped on to the stack.

Shed sides may be open (fig. 1.37) or slatted. Adjustable slats enable the airflow to be regulated to give greater control over the drying process. End protection can also be provided by this method – unprotected board ends are liable to split as a result of the ends drying out before the bulk of the timber, hardwoods like oak and beech are particularly prone to this problem. Other methods used to resist this particular seasoning defect are shown in figure 1.38, namely:

● treating the end grain with a moisture-proof sealer – for example, bituminous paint or wax emulsion, etc.;
● nailing laths over the end grain – thick laths should be nailed only in the middle of the board, to allow movement to take place;
● hanging a drape over the end of the boule or stack.

b Site conditions

As previously stated the whole site should be well drained, kept free from vegetation by blinding it with a covering of ash or concrete, and kept tidy.

If fungal or insect attack is to be discouraged, 'short ends' and spent piling sticks should not be left lying around.

Sheds should be sited with enough room left for loading, unloading, carrying out routine checks, and other operations.

c Stacking the timber (fig 1.39)

The length of the stack will be unlimited (depending on the timber lengths), but its height must be predetermined to ensure stability, and the stack must be built to withstand wind. The width should not exceed 2 metres, otherwise crossed airflow may well be restricted to only one part of the stack, however, adjacent stacks can be as close to each other as 300 mm.

d Piling sticks (stickers)

Piling sticks (stickers) should never be made from hardwood, or they could leave dark marks across the boards (fig 1.52). Their size and distance apart will vary, according to board thickness, drying rate, and species. They must always be positioned vertically one above the other, otherwise boards may 'bow' as shown in figure 1.39(a). Stacks with boards of random length may require an extra short stick as shown in figure 1.39 (b).

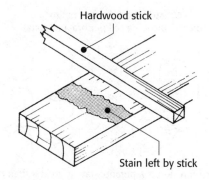

Hardwood stick

Stain left by stick

Fig 1.52   Stick marks (staining)

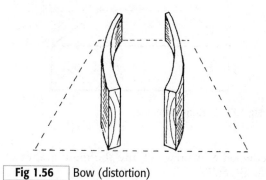

Fig 1.56   Bow (distortion)

Fig 1.53   Cupping (distortion)

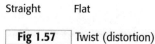

Straight   Flat

Fig 1.57   Twist (distortion)

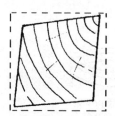

Fig 1.54   Diamonding (distortion)

Fig 1.58   Collapse (washboarding) (distortion)

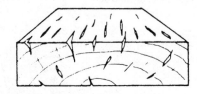

Fig 1.59   Surface and end checking

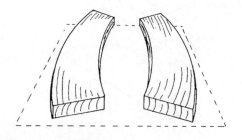

Fig 1.55   Spring (distortion)

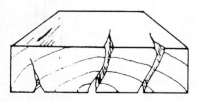

Fig 1.60   End splitting

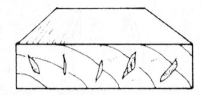

Fig 1.61   Honeycombing

Redwood & whitewood into the main grades of 'A', 'B', & 'C'.

Grade 'A' (the highest grade), is sub-divided into grades A1 (best quality) to A4. Grades B & C and not sub-divided.

As a comparison to their old system of classification:

CURRENT GRADE		PREVIOUS GRADE	
Grade A	=	US (unsorted) made up of	Firsts (1st) to Fourths (iv)
B	=	Fifths (v)	
C	=	Sixths (vi)	

Grades can differ across borders, and the shipping ports from which timber is dispatched. Identity of the port can be seen on the cut ends of timber as shown in figure 1.65 – these marks

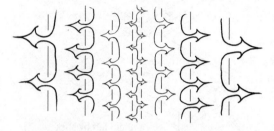

Moisture being drawn out to outer surfaces via cell perforations (pits)

NOTE: If moisture evaporation rate is greater than its movement through the wood from its centre then an imbalance will result

Fig 1.63   Moisture movement through wood

are known as *shipping marks* and are a very useful guide when buying a set type and quality of timber. For example, the shipping mark shown in figure 1.65 would indicate that the timber originated from Sweden and was shipped from Gothenburg and that it is of best quality unsorted (US) softwood.

Canadian grading rules will apply mainly to timber imported from North America, particu-

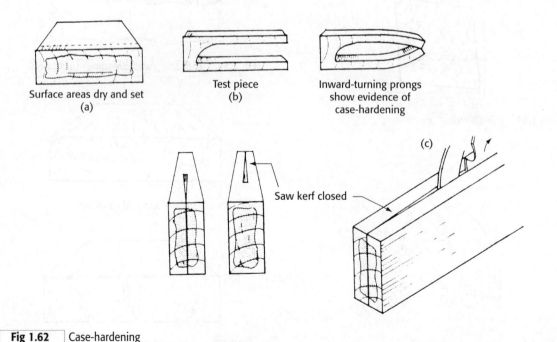

Surface areas dry and set
(a)

Test piece
(b)

Inward-turning prongs show evidence of case-hardening

(c)

Saw kerf closed

Fig 1.62   Case-hardening

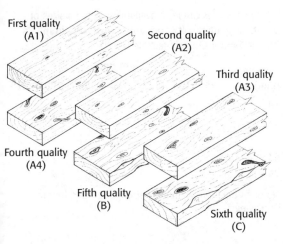

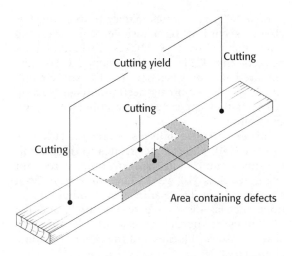

Fig 1.64   How appearance grades of European softwood might be seen – general guide only

Fig 1.66   Area's of clear yield of a hardwood board

But we can assume that the best European grades of 'firsts' are compatible with Canadian grades of '*clear*', and like wise '*sixths*' with '*common*'.

Other areas of America and Brazil use different grading terms.

1.8.2 Commercial grading of hardwood

Grading rules will vary between countries. The National Hardwood Lumber Associating (NHLA) grading rules cover the USA, certain regions may however vary these rules. Malaysia has its own grading rules but uses similar grading principals to determine usable pieces of a board.

NHLA grading rules are based on the amount (yield) of clear wood that can be cut from a board, (cuttings [fig 1.66] – percentage of board clear of defects) for example:

The best grades of FAS (firsts & seconds) should yield between 83.33% to 100% of clear wood cuttings, the size of these cuttings are restricted. Whereas the worst grades of say, No. 3 Common Grade need only yield 33.33% to 25% of clear cuttings.

As for decorative boards, many of the characteristic features of the wood will often be found outside the clear cuttings yield – possibly not in the most expensive grades.

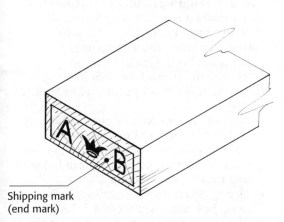

Shipping mark (end mark)

Note: unless otherwise stated the shipping marks will be coloured 'red'

Fig 1.65   Shipping marks can be a useful guide to species and quality

larly from the West Coast of Canada. Their standard for best quality is the term '*clear*' as a basis for timber that is free of any visual defects, and the term '*common*' for timber of a lower standard.

We cannot directly compare European grades with Canadian grades of:

- Clear
- Select merchantable
- Merchantable
- Common

1.8.3 Joinery timber

Whenever timber is incorporated as an item of joinery, a distinction is made as to which faces will be visible after installation.

BS EN 942 separates joinery timber into five classes according to allowable knot size, and other factors such as, shakes, resin and bark pockets (fig 1.24) – see section 1.6.10, discoloured sapwood (section 1.8.1) exposed pith (fig 1.5), and Ambrosia beetle damage (section 2.3.1) as they appear on the visible face of joinery timber.

Joinery timber classes – are: J2, J10, J30, J40, and J50. The number relates to the permitted maximum knot sizes in millimetres, for example, class J10 would have maximum knot sizes of 10 mm. There are however, other restrictions such as distance between knots, and the percentage of knots over the width of the timber, also, the size and location of the other degrading characteristics.

As an example, classes J2 – J30 would be used for constructing good quality joinery. Classes J40 – J50 for general purpose joinery.

Moisture content (section 1.7) should be as stated in table 1.6.

Durability (section 1.11e) – timber to be used for external joinery should either have an appropriate level of natural durability or, suitably treated with a wood preservative (see section 3).

1.8.4 Strength grading softwood (structural grading)

When ever an item such as a beam (joist, lintel, purlin etc.) is used within a building, it will have to be capable of carrying or supporting a specific load.

We have seen how commercial grading rules apply when grading timber for appearance. In this case, (irrespective of species) grading rules will be based on the ability of timber to withstand internal stresses which could be brought about by external forces. Therefore, every piece of timber that may be put into a situation of risk with regard to it's potential strength must be tested for its physical properties – then placed into a particular strength class and labelled accordingly.

This type of testing is called strength grading. Strength grading may be carried out at the saw Mill after conversion and drying (examples of possible strength graded timber sections are shown in table 1.8) or undertaken under factory conditions at the place of manufacture of structural components, such as a roof trusses.

There are two types of strength grading:

Table 1.8   A Guide to availability of sawn sizes of pre-stress graded softwood timber (extracted from 'Nordic Timber Council' fact file No. 5)

Thickness (mm)	Width (mm)						
	75	100	125	150	175	200	225
38	×	×	×	×	×	×	×
47	×	×	×	×	×	×	×
50*	×	×	×	×	×	×	×
63				×	×	×	×
75				×	×	×	×

Note: * Check on availability strength graded before specifying

1   visual grading, and
2   machine grading

a   **Visual strength grading of softwood** – visual-grading is a skilled operation, the operative must be qualified and hold a current certificate from an approved certification body. Graders have to determine the strength of each piece of timber handled, by inspecting all its surfaces for any:

- *Natural inherent defects* – such as type, size and distribution and number of Knots, wane, growth rate (growth rings), growth pattern (sloping grain etc) worm holes and rot etc. – see section 1.6
- *Drying defects* – such as splits, checks, twist, bow etc. – see section 1.7.8
- *Non-visible factors* – such as that density of the wood (relative weight – see section 1.11b) and moisture content – see section 1.7.1

Before rejecting it or classifying it into one of two grades;

1   *General structural* coded as GS
2   *Special structural* coded as SS

NB. grading SS. if the higher of the two grades

Figure 1.67(a) shows how markings may have appeared on the face of a stress graded timber prior to 1997.

Figure 1.67(b) shows how a suitable grade may now appear when graded within the UK.

Both visual strength graded timber from Canada under National lumbar Grades Authority (NLGA) rules, and USA National

Table 1.12   Relationship between wood cell distribution, type and function

Distribution	Tissue type	Location	Function
Axial cells	Tracheids (figs. 1.78 & 1.81)	Mainly softwood, some hardwoods (e.g. oak)	Provide strength and conduct sap
Axial and radial cells	Parenchyma (figs. 1.83 & 1.85 b & c)	Softwoods and hardwoods	Conduct and store food
Axial cells	Fibres (figs. 1.79, 1.80 & 1.85 a & c)	Hardwoods	Provide strength
Axial cells	Vessels or pores (figs. 1.79, 1.80 & 1.86 b & ?)	Hardwoods	Conduct sap

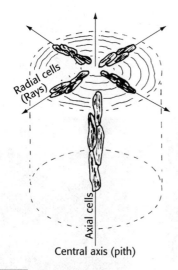

Fig 1.76   Wood cell distribution

pattern. Trees grown in tropical regions may not produce these rings annually, or may even not produce bands at all, because the climate responsible for growth may be such that continuous growth occurs, the only changes being those which reflect a wet or a dry season.

Ageing may produce another unmistakable feature – heartwood (duramen) and sapwood (alburnum). Heartwood is usually (but not always) darker in colour, due to chemical changes, inactive or dead tissue. It strengthens the tree by acting as its spine. The amount of heartwood will depend on the tree's age – the older the tree, the greater the amount. Sapwood (secondary xylem) is the active part of the tree, where cells conduct sap and store food. Young trees contain mostly sapwood tissue. Figure 1.77 is the key which shows that from the smallest sample of wood it is

possible – with the aid of a microscope – to see how the intricate wood structure is formed.

1.10.1 Structure of softwood (figs 1.77 and 1.78 to 1.81)

Softwoods belong to a group of plants called '*Gymnosperms*' (usually cone-bearing with naked seeds). The bulk of their 'stem' tissue is made up of axial tracheids – see table 1.12.

Tracheids (fig. 1.81) are elongated pointed cells, usually 2 to 5 mm long. Their interlaced arrangement provides the stem with its support. Holes in the cell walls allow sap to percolate from cell to cell. These perforations are known as 'pits', and there are two types: 'bordered' and 'simple'. Figure 1.81 and 82 show how tracheids use bordered pits to either allow or stop the flow of sap. Injury or ageing (sapwood becoming heartwood) will result in the closure of these pits – which could explain why preservative penetration is often difficult within heartwood.

Softwood 'rays' – responsible for food storage, among other things – generally consist of a line of single radial 'parenchyma' cells (fig. 1.83(a)), stacked in groups one upon the other and interconnected by 'simple' pits (fig. 1.84). In some softwood species, ray parenchymas may be accompanied by radial tracheids and resin canals (horizontal) and by vertical resin ducts. A vertical resin duct is shown in the transverse section in fig. 1.78.

1.10.2 Structure of hardwood (figs 1.79, 1.80, and 1.85)

Hardwoods belong to a group of plants called 'Angiosperms' (broad-leaved plants with enclosed

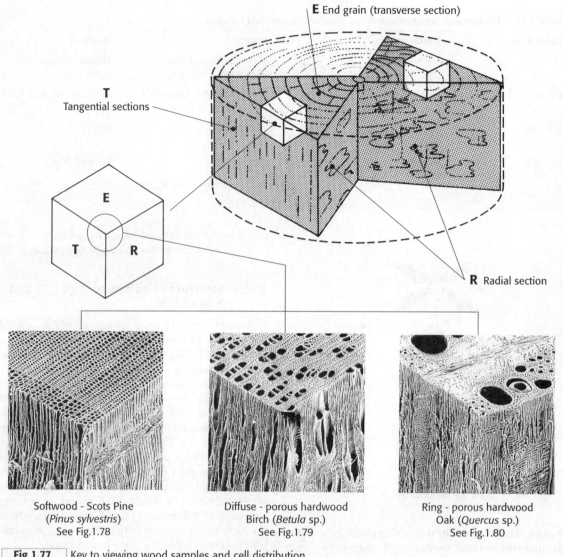

E End grain (transverse section)

T Tangential sections

E

T   R

R Radial section

Softwood - Scots Pine
(*Pinus sylvestris*)
See Fig.1.78

Diffuse - porous hardwood
Birch (*Betula* sp.)
See Fig.1.79

Ring - porous hardwood
Oak (*Quercus* sp.)
See Fig.1.80

Fig 1.77 Key to viewing wood samples and cell distribution

seeds) – they have a more complicated structure than softwoods, with a wider range of cell formations. The axial cells in the main are 'fibres' (fig 1.85a), with simple pits (fig. 1.84). Fibres give the tree its strength. 'Vessels'. (fig. 1.85(b)) – often also called 'pores' – are responsible for conducting sap and have both bordered and simple pits. As a means of distinguishing the difference between the term's 'vessel' and 'pore' the following explanation may be helpful:

The term 'vessel' may be used when the cell is cut through longitudinally and exposed across either a tangential or radial section (fig's 1.79, 1.80 and 1.85). Conversely when it is exposed

across a transverse section (end grain), (fig. 1.79E and 1.80E), it may be referred to as a 'pore'. In either case they are one and the same.

Rays in hardwood, often form a very distinct feature of the wood. Unlike in softwood, ray tissue can be several cells wide as shown in fig. 1.83(b).

Size and distribution of vessels are particularly noticeable in hardwoods with distinct growth rings. Two groups of these are as follows.

- *Diffuse-porous hardwoods* (fig. 1.79) These have vessels of a similar diameter and more or less evenly distributed around and across

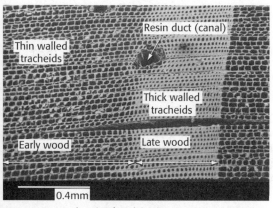

Transverse section (end grain) 'E'

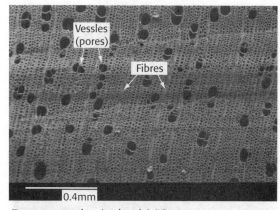

Transverse section (end grain) 'E'

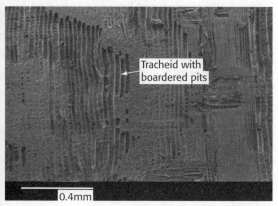

Radial section 'R'

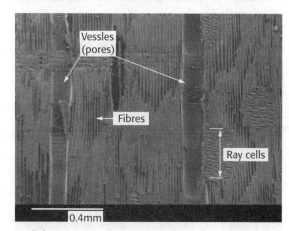

Radial section 'R'

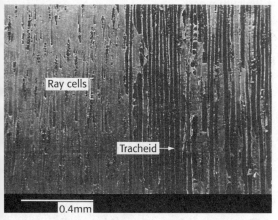

Tangential section 'T'

Fig 1.78 Softwood – scots pine (Pinus sylvestris)

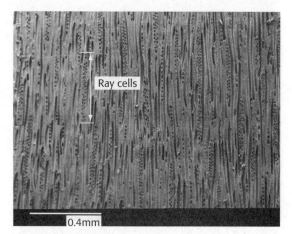

Tangential section 'T'

Fig 1.79 Hardwood – diffuse-porous – Birch (Betula spp)

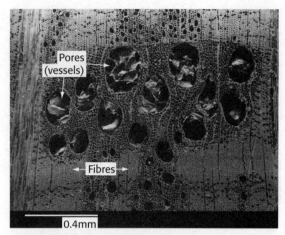

Transverse section (end grain) 'E'

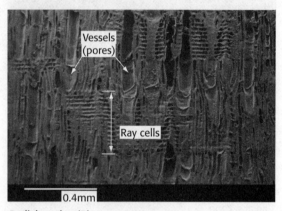

Radial section 'R'

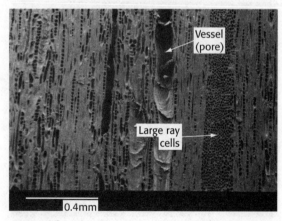

Tangential section 'T'

Fig 1.80   Hardwood – ring-porous – Oak (Quercus spp)

the growth-ring bands. Examples include beech, birch, sycamore, and most tropical hardwoods.

● *Ring-porous hardwoods* (fig. 1.80) In these, vessels are large in the earlywood, then become smaller with the latewood. Examples include ash, elm, and oak.

Hardwood can therefore be classed as 'porous' wood. (Note: 'porous' is a botanical term only.) Not only do the pores of a hardwood provide an important distinguishing feature between the two types, the lack of them would indicate a softwood.

1.10.3 Gross features of wood

Features like those listed below are often apparent without the use of magnification techniques! Simply by using our natural senses such as, sight, touch and possibly smell.

● growth rings
● sapwood & heartwood
● rays
● colour
● lustre
● odour
● cell intrusion (extractives)
● grain
● texture
● figure
● compression wood
● tension wood

a **Growth rings** – see sections 1.2.4 and 1.6.5.
b **Sapwood & heartwood** (also see sections 1.2.7 & 1.2.8) – if you were to see a transverse section (end grain) through a tree you may notice a demarcation line between the sapwood (outer bands) and heartwood (inner bands) – the amount of heartwood will generally depend on the age of the tree. In some species, sapwood can be lighter in colour than heartwood. There are however several exceptions to this rule which can help with initial identification. For example, European whitewood would show little difference in colour between its sapwood and heartwood, just like the rest of Spruce species, whereas European Redwood on the other hand, would show a distinct difference between the two, just like rest of the common named Pines.

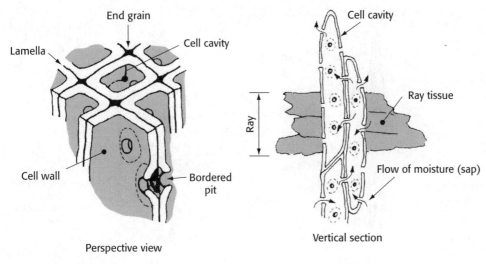

Fig 1.81   Tracheids

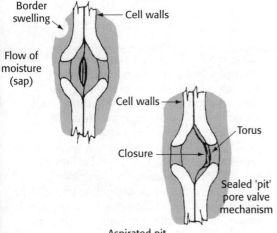

Fig 1.82   Vertical section through a bordered pit

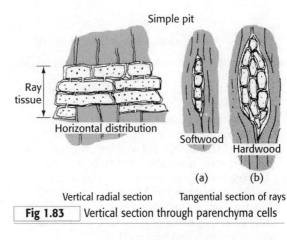

Fig 1.83   Vertical section through parenchyma cells

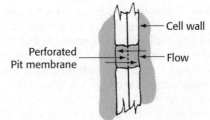

N.B. No border or swelling

Fig 1.84   Vertical section through a simple pit

c **Rays** (see section 1.2.5) – form a recognisable feature known as figure which will show up on the surface of the timber as the results of large and/or broad rayed species that have been quarter sawn, for example Oak which as shown in Figure 1.80 could produce a 'Silver figure'

d **Colour** – usually refers to the heartwood, as sapwood is generally pale in colour and in many cases insignificant. Table 1.14 and 1.15 gives examples of the heartwood colour of many popular wood species. This table should only be used as a general guide, because, the natural colour of the wood will change dramatically over period of time

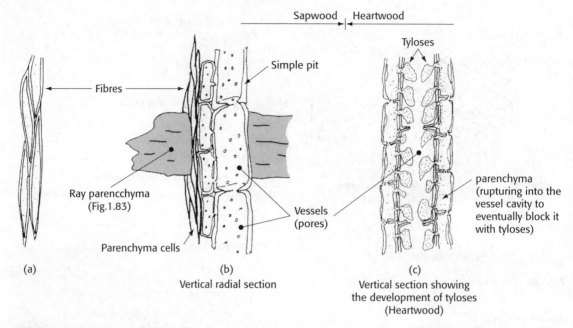

Fig 1.85   Vertical section through hardwood cells

particularly after lengthy exposure to air, light, or heat, this can have the effect of either darkening or even lightning the colour. The colour region of heartwood is brought about by the presence of extractives (g) retained within the cell and cell walls during its transition from sapwood to heartwood.

e **Lustre** – some wood surfaces when exposed to light take on a bright shiny appearance (lustre), whilst others may appear dull, both of which can be used to contrast one with the other as wood figure (J)

f **Odour** – all wood has its own particular odour some stronger than others, wood odour is usually more apparent whilst the wood is being cut – particularly by machine. Most odours tend to lessen with age, but somehow others don't. For example the odour given off by resinous pines are familiar to most wood workers, but I find most other wood species, with the possible exception of some hardwoods hard to identify by odour. Therefore it is best for each person to memorise his or her own interpretation of a particular odour.

g **Cell intrusion (extractives)** – Once wood cells have been cut open, their exposed cavities may have undesirable contents in the form of calcium salts or silica grains, as in

the case of some tropical hardwoods – for example, Iroko (calcium carbonate) and Keruing (silica). Silica grains can have a serious effect on the edge of cutting tools. Organic substances such as 'gums' in hardwood or 'resins' in softwood may also be present. Other substances such as oils, acids, tannins, and latex are also to be found in wood. These are known as 'extractives'.

Woods such as oak, chestnut, Douglas fir, and Western red cedar contain acids, which under moist conditions can corrode some metals. Woods containing tannins under similar conditions will react with ferrous (iron-containing) metals, the wood being stained a bluish-black colour.

h **Grain** – 'Grain' refers to the direction of the main elements of the wood. The manner in which grain appears will depend upon one or more of the following:
 ● the direction of the cut,
 ● the location of the cut,
 ● the condition of the wood,
 ● the arrangement of the wood cells.
Table 1.13 lists some common grain terms, together with a broad explanation of why they are so named.

i **Texture** – Texture is a surface condition resulting from the size and distribution of

Table 1.13   Grain terms and conditions

Grain terms and conditions	Explanation	Texture	Example of species	Remarks
End grain	Cross-cut exposure of axial and radial cell tissue	–	All	See Fig. 1.77
Straight grain	Grain which generally follows a longitudinal axial course	–	Keruing Kapur	Hardwood and softwood
Cross grain	Grain which deviates considerably from being parallel to the edge of the timber	–	Elm	Hardwood and softwood
Open or coarse grain	Exposed large vessels, wide rays, and very wide growth rings	Coarse	Oak, Ash	Associated with texture
Close or fine grain	Exposed small vessels, narrow rays and/or narrow growth rings	Fine	Sycamore Softwoods	Associated with texture
Even grain	Generally uniform, with little or no contrast between earlywood and latewood	Even	Spruce	Associated with texture
Uneven grain	Grain elements vary in size and uniformity – distinct contrast in growth zone	Uneven	Douglas fir	Associated with texture
Curly or wavy grain	Direction of grain constantly changing	Uneven	Walnut	Rippled effect
Interlocking grain	Successive growth layers of grain inclined to grow in opposite directions		African mahogany Afrormosia, Sapele	Striped or ribboned figure
Spiral grain	Grain follows a spiral direction around the stem from roots to crown throughout its growth			Defect in timber, affecting structural use (often visible as checks in de-barked poles)
Sloping grain (diagonal grain)	A conversion defect resulting from straight-grained wood being cut across its natural axial growth pattern, or a growth defect resulting from an abnormality in an otherwise straight tree		Can occur in any species	Defect in timber, affecting structural use

Note: In addition to the above 'short grain' may result from timber being cut and may easily split due to the short length of its elements (e.g. within a trench sawn for a housing joint)

wood cells. Texture is usually associated with touch, but, unless the grain is filled, many surface finishes, especially high-gloss polishes will reveal the texture direction and intensity of light being the all-important factors.

Texture is directly related to grain condition, as shown in table 1.13, and is a typical characteristic of timber (tables 1.14 and 1.15).

j   **Figure** – Figure is best described as the pattern or marking which is formed on the surface of processed timber as a result of

wood tissue being cut through. For example, quarter-sawn oak exposes broad rays, producing what is known as 'silver figure' (see fig. 1.86a); the interlocking grain of African mahogany will reveal a 'stripe' or 'ribbon figure' (fig 1.86b). Tangential-sawn softwood, like Douglas fir, can show a very distinctive 'flame like' figure (fig. 1.86c).

Timber possessing these characteristics can be regarded as having natural decorative properties

Table 1.14 A guide to some properties of the heartwood of various softwood species

Softwoods:

Common name	Species Latin (botanical) name	Origin	Colour of heartwood	(a) Moisture movement	(b) Approx density (kg/m³) @ 15% mc	(c) Texture (see table 1.13)	(d) Working qualities (cutting – nailing, etc)	(e) Natural durability (heartwood resistance to fungi)	(f) Treatability heartwood	General usage
Canadian Spruce	Picea Spp	Canada	White/pale yellow	Small	400 to 500	Medium	Good	Not durable – slightly durable	Difficult	Construction work
Douglas fir	Pseudotsuga menziesii	N America & UK	Light reddish brown	Small	530	Medium (straight grained)	Good	Moderately durable – durable	Extremely difficult	Plywood. Construction work & Int. & ext., Joinery
Lodgepole pine	Pinus contorta	N America	Yellow/pale brown/red tinge	Small	470	Fine	Good	Slightly to Moderately durable	Difficult to Extremely difficult	Construction work & Joinery
Parana pine	Araucaria angustifolia	S America	Pale brown/red streaks	Medium	550	Fine	Good	Not durable – slightly durable	Moderately easy	Plywood. Interior joinery
Radiata pine	Pinus radiata	S America S Africa Australia New Zealand	Yellow/pale brown	Medium	480	Medium	Good	Not durable – slightly durable	Moderately easy to difficult	Construction work and furniture
Redwood (European) Scots pine	Pinus sylvestris	Scandinavia/USSR UK	Pale yellow/brown to reddish brown	Medium	510	Medium	Good	Slightly to Moderately durable	Difficult to Extremely difficult	Construction work. Joinery and furniture
Sitca spruce	Picea sitchensis	N America UK	Pinkish/brown	Small	450	Course	Good	Not durable – slightly durable	Difficult	Construction work
Southern pine	Pinus Spp	Southern USA	Pale yellow/light brown	Medium	590	Medium	Medium	Slightly to Moderately durable	Difficult to Extremely difficult	Plywood. Construction & Joinery
Western Hemlock	Tsuga heterophylla	N America	Pale brown	Small	500	Fine	Good	Slightly durable	Moderately easy to difficult	Plywood. Construction work & Joinery
Western red cedar	Thuja plicata	N America	Reddish/brown	Small	390	Course	Good	Durable	Difficult to Extremely difficult	Exterior cladding. shingles/shakes
Whitewood (European)	Picea abies & Abies alba	Europe – USSR	White – pale yellowish brown	Medium	470	Medium	Good	Slightly durable	Difficult to Extremely difficult	Construction work & Joinery

Table 1.15 A guide to some properties of the heartwood of various hardwood species

Hardwoods:

Common name	Species Latin (botanical) name	Origin	Colour of heartwood	(a) Moisture movement	(b) Approx density (kg/m³) @ 15% mc	(c) Texture (see table 1.13)	(d) Working qualities (cutting – nailing, etc)	(e) Natural durability (heartwood resistance to fungi)	(f) Treatability heartwood	General usage
*Afrormosia	Pericopsis elata	West Africa	light brown	small	710	Medium/ fine	medium	durable to very durable	extremely difficult	interior and exterior joinery. cladding
Ash (American)	Fraxinus spp	USA	grey to brown	medium	670	course	medium	not durable	easy	interior joinery – trim. Tool handles
Ash (European)	Fraxinus excelsior	Europe	white – light brown	medium	710	medium/ course	good	not durable	moderately to easy	interior joinery
Beech (European)	Fagus sylvatica	Europe	white to pale brown – pinkish if steamed	large	720	fine	good	not durable	easy (red heart difficult)	interior joinery – flooring – plywood
Birch (American)	Betula spp	North America	light to dock reddish brown	large	670 to 710	fine	good	not durable	easy to moderately easy	plywood. Flooring – furniture
Chestnut (sweet)	Castanea sativa	Europe	yellow to brown	large	560	medium	good	durable	extremely difficult	interior and exterior joinery – fencing
Elm (European)	Ulmas spp	Europe	light brown	medium	560	course	medium	slightly durable	moderately easy to difficult	furniture – coffins – cladding
Gaboon	Aucoumea klaineana	West Africa	pinkish brown	medium	430	medium	medium	slightly durable	difficult	plywood (veneer & core)
Iroko	Milicia excelsa	West Africa	yellow to brown	small	660	medium	medium to difficult	durable to very durable	extremely difficult	Exterior and interior joinery. Constructional work
Keruing	Dipterocarpus spp	South-East Asia	pink brown to dark brown	large to medium	740	medium	difficult	moderately durable	difficult	heavy and general construction
African mahogany	Khaya spp	West Africa	reddish to brown	small	530	medium	medium	moderately durable	extremely difficult	interior joinery
American mahogany	Swietenia macrophylla	central and South America	reddish to brown	small	560	medium	good	durable	extremely difficult	interior and exterior joinery

(continued)

Table 1.15 A guide to some properties of the heartwood of various hardwood species (*continued*)

Hardwoods (*continued*):

Common name	Species Latin (botanical) name	Origin	Colour of heartwood	(a) Moisture movement	(b) Approx density (kg/m³) @ 15% mc	(c) Texture (see table 1.13)	(d) Working qualities (cutting – nailing, etc)	(e) Natural durability (heartwood resistance to fungi)	(f) Treatability heartwood	General usage
Maple, rock (hard)	*Acer saccharum*	North America	creamy white	medium	740	fine	medium	slightly durable	difficult	flooring – furniture
Meranti, dark red (red lauan)	*Shorea spp*	South-East Asia	medium to dark red brown	small	710	medium	medium	slightly durable to durable	extremely difficult	interior and exterior joinery. Plywood
Meranti, light red (white lauan)	*Shorea spp*	South-East Asia	pale pink	small	550	medium	medium	slightly durable to moderate	extremely difficult	interior joinery. Plywood
Oak, American red	*Quercus spp*	North America	Yellow/brown – red tinged	medium	790	medium	medium	slightly durable	moderately easy to difficult	interior joinery
Oak, American white	*Quercus spp*	North America	pale yellow to mid-Brown	medium	770	medium	medium	moderately durable to durable	extremely difficult	interior and exterior joinery
Oak, European	*Quercus robur, or petraea*	Europe	Yellow/Brown	medium	720	medium to course	medium to difficult	durable	extremely difficult	interior exterior joinery. Construction. Fencing
Ramin	*Gonystylus spp*	South-East Asia	white to pale yellow	large	670	medium	medium	not durable	easy	Mouldings. Dowelling
Sapele	*Entandrophragma cylindricum*	West Africa	reddish brown with marked stripe	medium	640	medium	medium	moderately durable	difficult	interior joinery. Plywood veneer
Sycamore	*Acer pseudo-platanus*	Europe	white to yellowish white	medium	630	fine	good	not durable	easy	interior joinery
Teak	*Tectona grandis*	Burma. Thailand	golden to dark brown	small	660	medium	medium	very durable	extremely difficult	interior and exterior joinery. Garden furniture
Utile	*Entandrophragma utile*	West Africa	reddish brown	medium	660	medium	medium	moderately durable to durable	extremely difficult	interior and exterior joinery. Furniture

* Imports subject to export permit from country of origin

Quarter sawn species with broad rays e.g. Oak

Quarter sawn species with interlocking grain e.g. African Mahogany

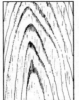
Plain or tangenital sawn species with distinct growth rings e.g. Redwood or Douglas Fir or ring porous hardwoods

Silver figure
(a)

Stripe or ribbon figure
(b)

Flower or flame figure
(c)

Fig 1.86 Decorative figure

k Compression wood (softwoods) – this condition is dealt wit in section looking at 'Reaction Wood' (section 1.6.1).

l **Tension wood** (hardwoods) – this condition is dealt within section looking at 'Reaction Wood' (section 1.6.1).

1.11 Properties of timber

The physical properties of timber will depend on the wood species, growth characteristics, subsequent conversion, and final processing.

Table 1.14 and 1.15 form the basis for selecting which of the hardwood and softwood species listed will be suited to a given situation, with regard to their:

- moisture movement,
- density,
- texture,
- working qualities,
- durability,
- general usage

a **Moisture movement** – (Column (a)) refers to the amount of movement which might affect the dimensions and shape if timber cut from that particular species is subjected to conditions liable to alter its moisture content (m.c.) after having been dried to suit its end use. i.e. in a state of equilibrium with its environment. Any increase or decrease in moisture content will result in the timber either expanding (swelling) or contracting (shrinking). Proportional amounts of change are illustrated in fig. 1.33 and table 1.6.

b **Density** – (Column (b)) refers to the mass of wood tissue and other substances contained within a unit volume of timber. It is expressed in kilograms per cubic metre (kg/m³). It therefore follows that the m.c. of timber must affect its density – figures quoted in tables 1.14 and 1.15 are average for samples at 15% m.c.

Some factors affecting timber density (fig. 1.87) are:

- bound water within cell walls (bound moisture),
- the presence of free water within cell voids (free moisture),
- the presence or absence of extractives
- the amount of cell-tissue in relation to air space.

Because the density of solid wood tissue is the same for all species (1506 kg/M³), the differences in density between the species listed in table 1.14 and 1.15 indicate the proportion of cell tissue per unit volume.

Density is a property closely associated

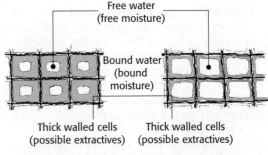
Free water (free moisture)

Bound water (bound moisture)

Thick walled cells (possible extractives)

Thick walled cells (possible extractives)

Horizontal section through thick walled cells

Horizontal section through thin walled cells

Fig 1.87 Factors that can effect wood density

with the hardness of timber and its strength properties (see 'Strength' below at 'g').

c **Texture** – (Column (c)) – as previously stated (1.10.3 i) – refers to grain character and condition, which can greatly influence the use and working properties of timber.

d **Working qualities** – (Column (d)) refers very broadly to how the timber will respond to being cut and machined, but not necessarily to how it will respond to glue. Factors which may influence these qualities include grain condition (table 1.13); a hard, soft, or brittle nature; and the presence of any cell intrusions (extractives) which could cause the wood to be of an abrasive, corrosive, or greasy nature.

e **Durability** – (Column (e)) in this case, depends on the wood's natural resistance to fungal attack, which is influenced by moisture content and the amount of sapwood.

The standards of durability quoted here refer to sample species of a 50 mm × 50 mm section of 'heartwood' left in contact with the ground for a number of years. Species quoted as *'not durable'* may be regarded as having less than five years' life, whereas *'very durable'* could have more than 25 years' life.

There are five natural durability (resistance to wood-destroying fungi) classes:

Class 1.   Very durable
Class 2.   Durable
Class 3.   Moderately durable
Class 4.   Slightly durable
Class 5.   Not durable.

For a true durability classification you should refer to BS EN 350–1, and for relationship to hazard classes refer to BS EN 460.

f **Treatability (permiability)** – (Column (f)) – as can be seen from table 1.16. In this case treatability refers to ease whereby a wood preservative can be impregnated into heartwood when using standard vacuum pressure methods.

Restriction of movement can be due to many reasons, for example:
● Cell size and distribution
● Cell blockage ('pit' closures [aspirated – fig 1.82] – 'vessel' closures [development of 'tyloses' – fig 1.85(c)])
● Moisture content

● Sapwood in relation to heartwood
● Knots – size, number, and distribution
● Presence of intrusions (extractives)

In some case permeability can be increased by using rough sawn timber in stead of planed – in extreme cases a process known as *incising* may be employed. This involves passing timber through a special machine which cuts a series of small incisions into the surface of the timber to induce liquid to enter the wood.

Reference can also be made to BS EN 350–2

g **Strength** – Because of its high strength-to-weight ratio, timber can be suited to many situations requiring either compressive or tensile strength. Figure 1.88 shows how a

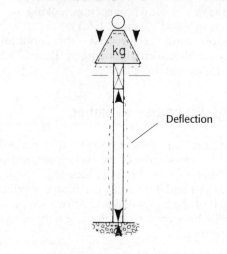

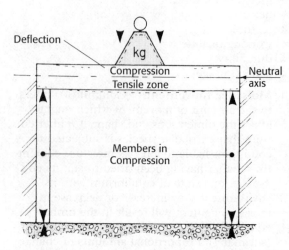

Fig 1.88   Structural timber being subjected to a load

Table 1.16   A guide to treatability (permeability) showing lateral penetration of preservative using standard pressure methods

Classification 'E' 'ME' 'D' 'ER'		Lateral penetration (heartwood)	Species – Examples – also see Tables 1.14 and 1.15
Easy (E)		No resistance or difficulty	Ash (American). Beech. Ramin. Sycamore.
Moderately Easy (ME)		6–18 mm after 2–3 hours	Parana pine. Ash (European). Oak (American Red).
Difficult (D) (difficult to treat)		Difficult to penetrate	Canadian Spruce. Redwood. Sitca Spruce. Sapele.
Extremely Difficult (ED) (extremely difficult to treat)		Minimum absorption	Douglas Fir. Whitewood (European). Afrormosia. African and American Mahogany. Meranti.

structural timber member can be stressed. The posts are being compressed from between their ends by a beam and its load and the ground. The beam is being subjected to stresses of compression (wood tissue being pushed together) on its upper surface and tension (tissue pulled apart) on its lower surface – the neutral axis being a hypothetical dividing line where stresses are zero.

If either of the stressed zones – particularly the tension zone – is damaged or distorted, either by natural defects such as knots etc. (section 1.6.9) or by defects associated with seasoning (section 1.7.8), the beam could be drastically weakened. The cutting of notches in stressed timber members must always be strictly controlled (see book 2) – notches should not be cut into the tension zone.

Structural weakness may also be due to other factors. Examples are given in table 1.17 portraying the following:
- timber section in relation to load;
- low density of timber (particularly of the same species) at an acceptable moisture content;
- high moisture content;
- direction of growth rings – wood cells and tissue should be positioned to provide greatest strength;
- fast growth of softwood produces low-density wood because of the large thin-walled cells (tracheids);
- slow growth of hardwood produces a greater number of large vascular cells (vessels) and a lesser number of strength-giving fibres.

h **Effects of fire** – timber is, of course, combustible. The importance of this is not that timber burns but for how long it retains its structural stability while being burnt. It would be impossible to light a camp fire with heavy logs – it is much better to start with dry twigs and then build up the size of material gradually until the fire has a good hold. Then, and only then, can the logs be added so that the fire will, provided there is enough draught and conditions are right, burn steadily for some time – probably hours without needing replenishment. It can therefore be said that the rate at which timber burns must be related to:

Table 1.17 Guide to the condition of timber in relation to its general strength characteristics

Condition	Stronger	Weaker
a Siting – position (stiffness)		
b Density	High (heavy)	Low (light)
c Moisture content	Low	High
d Direction of growth rings	Tangential-sawn softwood	Quarter-sawn softwood
e Fast growth	Ring-porous hardwood	Softwood
f Slow growth	Softwood	Ring-porous hardwood

- its sectional size,
- its moisture content,
- its density,
- an adequate supply of air.

Therefore, provided an adequate section is used, timber can retain its strength in a fire for longer than steel or aluminium, even though they are non-combustible.

One effect flaming has on timber, is to form a charcoal coating over its surface (this is measurable and known as its *charring rate*), which acts as a heat insulator, thus slowing down the rate of combustion. Failure rate therefore depends on the cross-sectional area of the timber.

Flame-retardant treatments are available but, although useful for unstressed wood, their presence may have an adverse effect on the strength of timber (see section 1.11g).

i **Thermal properties** – The cellular structure of timber provides it with good thermal-insulation qualities – species which are light in weight and of low density are particularly effective insulators, where structural strength is not important.

Examples of good practical thermal insulation and poor conductive properties, can be found within wall components of timber framed buildings. These help prevent heat loss from within, and cold intrusion from without, or vice versa, depending on the local environment or climate. Another example would be the use of wood handles for metal (good conductor of heat) cooking utensils etc.

j **Electrical resistance** – Although timber generally has good resisting properties against the flow of an electrical current, its resistivity will vary according to the amount of moisture it contains. Moisture is a good electrical conductor, and this is the basis on which moisture content is measured when using a moisture meter (see section 1.7.1b).

k **General usage** – the final columns of table 1.14 and 1.15 indicates the situations to which the particular species is best suited, provided the specimen is of the correct quality with regard to either strength or appearance. There are only a few species which satisfy both strength and appearance requirements, and those which do, are usually very expensive – English oak being a good example.

Table 1.18 should give you some idea of how we often refer to a timber section in

Table 1.18   Examples of timber sections and their possible end use

Timber section	Work stage	End use
 Sawn	Structural carpentry	Carcassing – skeleton framework (walls, floors, roofs)
 Sawn and processed	Carpentry – 1st fixing (work done before plastering)	Partitions, floors, decking, grounds, window boards, door casing/linings
 Processed board and trim	Carpentry – 2nd fixing (work done after plastering)	Shelving, trims (skirting boards, architrave), falsework (panelling, pipe boxes, etc.)
 Processed boards, sections and trims	Joinery (Exterior) Joinery (interior)	Windows, doors, gates, etc. Carcase units, cupboards, doors, stairs, and fitments

relation to its end use. Some of these terms you may not be familiar with, such as:

- carcassing
- first fixing
- second fixing
- joinery (exterior)
- joinery (interior)

But by comparing the headings of *'work stage'* and *'end use'* a simple explanation will be found.

2

Enemies of Wood and Wood Based Products

If conditions are favourable, fungi or wood-boring insects or both can attack most wood. Such attacks are often responsible for the destruction of many of our trees (e.g. Dutch elm disease) and for the decomposition and subsequent failure of many timbers commonly used in building.

As shown in figure 2.1 (stages a–c), in the natural world the work of wood destroying fungi and insects is instrumental as natures way of biologically cleaning up the forest floor of all dead wood and other organic matter. The outcome of this action, is to produce as a bi-product nutrients and floor litter to help support various life forms as well as for the regeneration of trees.

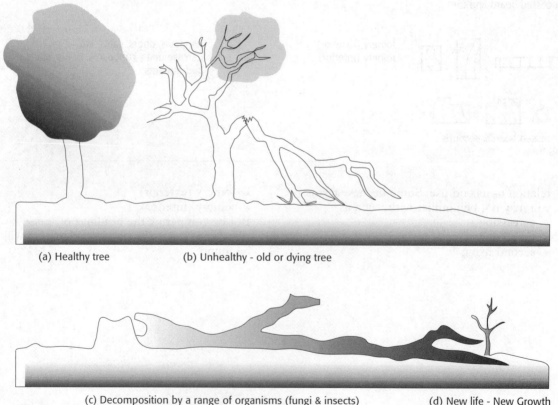

(a) Healthy tree   (b) Unhealthy - old or dying tree

(c) Decomposition by a range of organisms (fungi & insects)   (d) New life - New Growth

Fig 2.1   The natural process of decomposition and regeneration

Fungi – there are many different types and species, all belonging to the lower plant kingdom. The fungi that concern us are those that take their food from, and live on or in, growing trees and more importantly the timber cut from them.

Attack by Insects – like fungi there are many different types of wood boring insects each with their own special life style. There are those that prey on damaged living trees, recently felled trees, and those with a preference for different types and conditions of timber.

2.1 Non-rotting fungi (Sap-staining fungi)

These fungi, also known as *'blue-stain'* or *'blueing'* *fungi*, attack timber with a minimum moisture content of about 27%, but not saturated. They obtain their nourishment from the cell contents of sapwood, leave wood cell walls unscathed, therefore, the main degrading effect these fungi have on timber is, as shown in figure 2.2 the bluish to black discolouration they leave on the surface of processed timber.

Because these fungi only attack wood with high moisture content, it is necessary that the conversion and drying is carried out as early as possible after the trees have been felled.

2.2 Wood-rotting fungi

These fungi live off the cell walls of the wood, thus causing its whole structure to decompose and eventually collapse. Their growth requirements are simple, wood with above 20 % m.c. for initial germination, and subsequent nourishment in the form of non-durable sapwood.

Durable timbers are those which have a natural resistance to fungal attack (see section 1.11e); however, the majority of timbers have either to be kept permanently below 20% m.c. or undergo treatment with a suitable wood preservative (chapter 3), as an assurance against attack.

All wood-destroying fungi have a similar life cycle, a typical example is shown in figure 2.3. The spores have been transported from the parent plant (fruiting body or sporophore) – by wind, insects, animals, or an unsuspecting human – to a suitable piece of fertile wood where germination can take place. Once established, the fungus spreads its roots (hyphae) – which are in fact the body of the fungus – into and along the wood in search of food, eventually to become a mass of tubular threads which collectively are called *mycelium*, and which will eventually produce a fruiting body.

Fruiting bodies (sporophores) can take the shape of stalks, brackets, or plates etc. (fig. 2.4), depending on the species of fungus. Each fruiting body is capable of producing and shedding

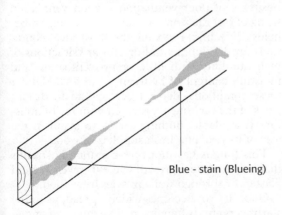

Fig 2.2 | The effect of a sap-staining fungus

Blue - stain (Blueing)

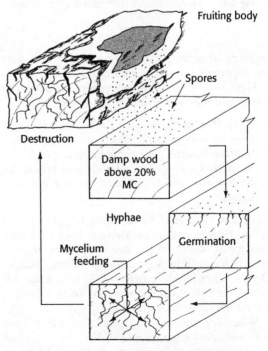

Fig 2.3 | The life cycle of wood destroying fungi

Fruiting body
Spores
Destruction
Damp wood above 20% MC
Hyphae
Germination
Mycelium feeding

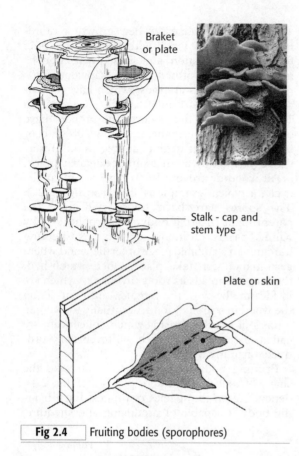

Braket or plate

Stalk - cap and stem type

Plate or skin

| **Fig 2.4** | Fruiting bodies (sporophores) |

millions of minute spores, of which only a very small proportion will germinate.

Here, we are mainly concerned with the two most common wood-destroying types:

- **'Dry Rot' fungus** (*Serpula lacrymans – formerly* known as *Merulius lacrymans*) and,
- **'Cellar fungus'** (*Coniophora puteana –* formerly *Coniophora cerebella)*, also known as *'cellar rot'* which is one of the **wet-rot fungi.**

Both Dry Rot and Cellar fungus belong to a group of rots known as *'brown rots'*, because the wood which they attack appears darker, brown in colour and, on drying, becomes brittle. On the other hand, *'white rots'*– which include other wet rots outside the scope of this book – lighten the colour of the wood.

Wherever the following conditions exist, fungi will inevitably become established – the type of fungus and its characteristic life style being in the main determined by the amount of moisture present in the wood.

Fungi need the following elements to become established:

- *Food* – in the form of cellulose from the woody tissue of sapwood and non-durable heartwood (see section 1.11e, 'Durability').
- *Moisture* – in the first instance, wood will have had to attain a moisture content in excess of 20%.
- *Temperature* – Low temperatures may reduce their growth; high temperatures on the other hand, will kill them.

The preferred temperature for *Serpula lacrymans* is in the region of 20°C.

- *Air* – an essential requirement for the growth and respiration of fungi.

Once fungi are established, it is only a matter of time before the wood substance starts to decompose and structural breakdown occurs, with the result, that the wood:

- loses its strength,
- becomes lighter in weight (reduced density),
- changes its colour by becoming darker (brown rot) or lighter (white rot) – depending on the type of fungus.
- becomes more prone to insect attack
- loses its natural smell – taking on a foisty or musty smell.

The following text should be read in conjunction with table 2.1.

2.2.1 **Dry-rot fungus** (*Serpula lacrymans*)

Timber becomes liable to dry-rot attack when its moisture content exceeds 20% and it is sited in positions of poor ventilation – good ventilation being one of the prime factors for keeping timber below 20% m.c. Probably the most ideal conditions for development of dry rot are situations of high humidity with little or no ventilation and moisture content of between 30 to 40%. Under these conditions, dry rot can spread to distant parts of the building – provided a source of moisture is available, nothing seems to stand in the way of its reaching fresh supplies of wood.

This fungus has moisture-conducting strands that help it sustain growth while it travels behind plaster, or through walls in search for food.

Once dry rot becomes widely spread, its initial starting point (often its main source of moisture) becomes more difficult to find – the fungus may have travelled from room to room, floor to

Table 2.1   General characteristics of wood destroying fungi: Dry Rot and Cellar fungus (Cellar Rot)

	Name of fungus	
	Dry rot **(Serpula lacrymans),** **Figs 2.6 to 2.8**	**Cellar fungus (Cellar Rot)** **(Coniophora puteana),** **Figs 2.10 and 2.11**
Appearance:		
fruiting body (sporophore)	Plate or bracket, white-edged with rust-red centre (red spore dust)	Plate – not often found in buildings – olive green or brown
mycelium	Fluffy or matted, white to grey, sometimes with tinges of lilac and yellow	Rare
strands	Thick grey strands – can conduct moisture from and through masonry.	Thin brown or black strands, visible on the surface of timber and masonry
How wood is affected	Becomes dry and brittle, breaking up into large and small cubes, brown in colour,	Exposed surfaces may initially remain intact. Internal cracking, with the grain and to a lesser extent across it. Dark brown in colour.
Occurrence	Within buildings, originating from damp locations	Very damp parts of a building, e.g. affected by permanent rising damp, leaking water pipes, etc.
Other remarks	Once established it can spread to other, drier, parts of the building.	Fungal attack ceases once dampness is permanently removed.

Note: Both the above fungi can render structural timbers unsafe.

floor, even into or from the roof, and have re-established itself with another moisture supply. Places that are most suspect will have evolved from bad design, bad workmanship, or lack of building maintenance. Some of these locations are shown in figure 2.5.

A typical outbreak of dry rot to a timber floor is shown in figures 2.6 and 2.7.

Tell-tale signs of an outbreak could be one or more of the following features:

- Smell – a distinct mushroom-like odour (damp and musty).
- Distorted wood surface – warped, sunken (concave), and/or with shrinkage cracks (fig.2.6). Tapping with a hammer produces a hollow sound, and the wood offers no resistance when pierced with a knife.
- The appearance of fruiting bodies (sporophores) in the form of a 'plate' (skin, fig 2.4) or 'bracket'.
- The presence of fine rust-red dust, which is the spores from a fruiting body.

Exposed unpainted timber may reveal a covering of a soft carpet of whitish mycelium. Established mycelium may appear as a greyish skin with tinges of yellow and lilac. Wood already under attack will have contracted into cuboidal sections (fig. 2.8).

The fruiting body of the fungus, which is capable of producing millions of spores, appears as an irregularly shaped rusty-red fleshy overlay with white edges.

Eradication of dry rot

The following stages are necessary even for the smallest of outbreaks.

i   Eliminate all obvious sources of dampness – some possible causes are shown in figure 2.5.

ii   Investigate the extent of the outbreak – removing woodwork and cutting away plasterwork as necessary.

iii   Search for further causes of dampness both within and outside the areas of attack.

iv   Remove *all* affected woodwork, (timber should be cut back as shown in figure 2.9) and fungus from the building, to where it can be safely incinerated or disposed of. *In both these cases the method and type of disposal will generally require local authority approval*

v   Surrounding walls, concrete floors, etc. may require sterilising and treatment with a suitable fungicide.

vi   All replacement timber must be below 20% moisture content.

vii   In areas were there is a risk of re-infestation, replacement timber will require preservative treatment (with special attention being given to end grain).

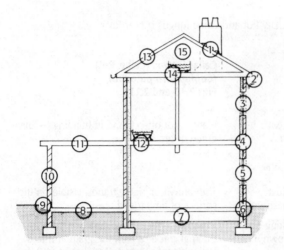

Fig 2.5   Possible causes of dampness
1 – defective step flashing
2 – defective or blocked gutter/fallpipe
3 – window condensation, or insufficient external
    weathering
4 – bridged cavity wall (mortar droppings, etc.)
5 – defective or omitted vertical damp proof course
    (DPC) around doors and/or windows
6 – defective or omitted horizontal DPC, blocked air
    brick/grate, or ground level built-up above DPC
7 – no ventilation to under-floor space
8 – defective or omitted damp proof membrane (DPM)
9 – defective or omitted DPC, or ground level built-up
    above DPC
10 – solid wall of porous masonry
11 – defective roof covering, unvented rood void
    (cold-deck construction), defective or omitted
    vapour barrier
12 – defective plumbing/water spillage
13 – defective roof covering
14 – defective plumbing
15 – unvented roof space

Note: See also 'Preservatives' in chapter 3 (section 3.3).

2.2.2 **Cellar fungus** (Coniophora puteana)

Also known as **Cellar Rot**. The most favourable locations for this rot are those which can permanently provide wood with a moisture content of about 40 to 50%. Examples of where suitable conditions may be found are listed below (see also fig. 2.5):

● in damp cellars or other rooms below
  ground level;

Fig 2.6   Surface effect of a dry-rot attack (with kind permission of Rentokil)

Fig 2.7   Severe attack of dry-rot exposed (with kind permission of Rentokil)

● beneath leaking water pipes and radiators;
● behind and under sinks and baths, etc. – due
  to persistent over-spill and splashing;
● areas of heavy condensation – windows,
  walls, roofs;
● areas where water can creep (by capillarity)

Fig 2.8 Dry-rot damage showing the cuboidal effect (with kind permission of Rentokil)

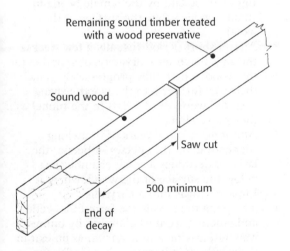

Fig 2.9 Cutting away timber affected by rot

Remaining sound timber treated with a wood preservative

Sound wood

Saw cut

500 minimum

End of decay

Fig 2.10 A typical wet-rot location (window sill) – with kind permission of Rentokil)

Fig 2.11 Wet-rot damage to end grain – (with kind permission of Rentokil)

and remain free from evaporation, behind damaged paintwork, under window sills (fig. 2.10), and thresholds, etc.

- above a defective damp-proof course (d.p.c.) or membrane (d.p.m.).
- timber in permanent contact with the ground, or sited below a d.p.c. or d.p.m.

However, unlike dry rot, once the moisture source has been removed and the wood has dried out, cellar rot becomes inactive. Figure 2.11 shows the effect of wet-rot damage on end grain.

The extent of the decay may not be obvious, at first because the outer surfaces of the timber

often appear sound. It is not until these surfaces are tested with the point of a knife or bradawl, that the full extent of the attack is known. The rot is, however, reasonably confined to the area of dampness.

One of the biggest dangers associated with cellar rot is that the area of its confinement may eventually become taken over by dry rot, in which case a major problem may well have developed.

Note: There are several other wet rots, all of which require reasonably high percentages of moisture. This can make formal identification difficult.

The term 'Soft Rot' is often used to describe softening of the surface layer of wood by a wet rot. This can occur when wood is, or has been in contact with the ground.

Eradication of wet rot (Cellar Fungus)

Unlike dry rot, the one controlling factor here is dampness. However, all wet rot should be treated in the following manner:

- Remedy as necessary any sources or defects responsible for the dampness (fig. 2.5).
- Dry out the building.
- Remove and safely dispose of affected timber, cutting back well into sound wood (fig. 2.9).
- Where, for example, there is a risk that the timber may not be kept dry, such as in contact with the ground, then a suitable wood preservatives should be applied to timber at risk (paying special attention to end grain); alternatively, use pre-treated timber – see section 3.4.1.
- All replacement timbers must be below 20% moisture content and treated with preservative (paying special attention to end grain).

Note: Because the strands of cellar rot do not penetrate masonry, treatment is more localised.

2.2.3 Preventing wood rot

Theoretically, rot should not occur in a building that has been correctly designed, built, and maintained. For prevention, all that is required is that the moisture content of wood should never be allowed to exceed 20%.

The only obvious places, (with the exception of garden fences, gates, and their posts), where wood is subject to conditions likely to achieve that level, are items of external joinery such as doors and windows, in each case, unless timber has an appropriate natural-durability rating it will require preservative treatment together with permanent protection against the entry of moisture.

2.3   Attack by wood boring insects

The term *wood boring insect or beetle* as opposed to woodworm can be confusing. Although the beetle is capable of biting holes into wood, it is the larva or grub (woodworm) of the beetle that is in the main responsible for the damage done to the wood. The endless tunnelling of the larva, while it feeds on the wood substance, brings about the damage. It is therefore inevitable that in extreme cases where mass infestation is present, there will always be the risk that the structure of the wood will be weakened.

There are many species of wood-boring insects, each with its own life style. For example, there are those which prey on living, recently felled, or dying trees, while others will only attack certain species or parts of them – and there are those which are not too particular. Table 2.2 lists those insects/beetles common to the UK.

General identification can often be made by one or more of the following characteristics:

a habitat,
b size and shape of the beetle,
c size and shape of the larva,
d size and shape of the bore dust (frass),
e size and shape of the exit (flight) holes,
f sound (death-watch beetle).

A typical life cycle of these insects (with the exception of the wood boring weevil) is as shown in figure 2.12 and described below:

1   Eggs will be laid by the female beetle in small cavities below the surface of the wood.
2   After a short period (usually a few weeks) the eggs hatch into larvae (grubs) and enter the wood, where they progressively gnaw their way further into the wood, leaving excreted wood dust (frass) in the tunnel as they move along.
3   After one or more years of tunnelling (depending on the species of beetle) the larva undermines a small chamber just below the surface of the wood, where it Pupates (turns into a chrysalis).
4   The pupa then takes the form of a beetle and emerges from its chamber by biting its way out, leaving a hole known as an exit or flight hole. (Collectively, these holes are usually the first sign of any insect attack.) The insect is now free to travel or fly at will – to mate and complete its life cycle.

Two insects not mentioned in table 2.2 are the Pin-hole borers (ambrosia beetles) and the wood wasp, which are associated with attacks on either the standing or newly felled tree.

2.3.1 Pin-hole borers (ambrosia beetles)

Attacks are noticeable via the black stained edges of their bore holes (tunnels), as shown in figure 2.13 and generally run across the grain of the wood, they can, depending on species be from 0.5 to 3 mm in diameter.

Table 2.2 General characteristics of some wood destroying insects

Name of insect	Common furniture beetle (*Anobium punctatum*), Fig. 1.48	Powderpost beetle (*Lyctus* spp.), Fig. 1.49	Death-watch beetle (*Xestobium rufovillosum*), Fig. 1.50	House longhorn beetle (*Hylotrupes bajulus*), Fig. 1.51	Wood-boring weevils (*Pentathrum* or *Euophryum* spp.), Fig. 1.52
Adult characteristics:					
size	3 to 5 mm	5 mm	6 to 8 mm	10 to 20 mm	3 to 5 mm
colour	Reddish to blackish brown	Reddish brown to black	Chocolate brown	Grey/black/brown	Reddish brown to black
flight	May to August	May to September	March to June	July to September	Any time
Where eggs laid on or in wood	Crevices, cracks flight holes, etc.,	Vessels of large-pored hardwoods	Fissures* in decayed wood, flight holes	Fissures* in softwood, sapwood	Usually on decayed Wood
Size of larvae	Up to 6 mm	Up to 6 mm	Up to 8 mm	Up to 30 mm	3 to 5 mm
Bore dust (frass)	Slightly gritty ellipsoidal pellets	Very fine powder, soft and silky	Bun-shaped pellets	Cylindrical pellets	Very small pellets
Dia. of flight (exit) hole	2 mm	0.75 to 2 mm	3 mm	6 to 10 mm Ellipsoidal(oval)	Up to 2 mm
Life cycle	2 years or more	10 months to 2 years	4 to 10 years	3 to 11 years	7 to 9 months
Wood attacked:					
type	SW and HW, plywood with natural adhesive	HW, with large pores e.g. oak, elm, obeche	HW, especially oak	SW	SW and HW, plywood with natural adhesive
condition	Mainly sapwood	Sapwood	Previously decayed by fungus	Sapwood first, heartwood later	Usually damp or decayed
location	Furniture, structural timbers etc.	Timber and plywood while drying or in storage	Roofs of old buildings, e.g. churches	Roofs and structural timbers	Area affected by fungus (usually wet rot) – cellars etc.
Other remarks	Accounts for about 75% of woodworm damage in UK. Resin-bonded plywood and fibreboard immune.	Softwood and heartwood immune	Attacks started in HW can spread to SW	Restricted to areas of Surrey and Berkshire in UK.	Attack continues after wood has dried out.

* Fissures – cracks, narrow openings, and crevices

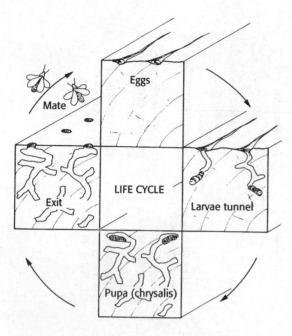

Fig 2.12   Life cycle of common wood boring insects (except wood boring weevils)

Tunnelled galleries at right angles to the grain lined with Ambrosia fungus

Sample of tropical hardwood such as mahogany, lauan, or meranti during or after an infestation of Pin-hole borers

Fig 2.13   Pin-hole borer damage – notice how in the main tunnelling is carried out across the grain

Tunnels become lined with a fungus, known as 'Ambrosia' Fungus, (hence the reason for using the term 'Ambrosia beetle') spores of which are introduced by the adult beetle thereby providing food for both the beetle and its larvae. When the moisture content of the wood is reduced to about 30% m.c., the fungus dies, and consequently so does the beetle as the food supply runs out, leaving behind the non-active black lined bore holes we associate with some tropical hardwoods.

2.3.2 **Wood-Wasp**

The wood wasp can produce a large bore hole of up to 9 mm in diameter. Trees most susceptible to attack by wood-wasps are usually unhealthy softwood trees, and as a result, timber derived from them may contain active larvae. However, any emerging wood-wasp would be unable to reinfest the wood once it had been dried.

The adult wood-wasp cannot sting, and is harmless to humans, although at first sight, because of its size (10 to 50 mm long), yellow and black/blue markings, long spine and loud buzzing sound, it is often mistaken for a hornet or similar species of wasp.

2.3.3 **Common furniture beetle** (Anobium punctatum – Fig 2.14)

The best known wood boring insect in the United Kingdom – more commonly known as *woodworm*. The larvae (woodworm) of this beetle are responsible for about 75% of the damage caused by insects to property and their contents – attacking both softwood and temperate hardwood, but with a preference for sapwood.

The adults range from reddish to blackish brown in colour, 3 to 5 mm in length, and live for about 30 days. The female lays eggs, about 80 in number, in small fissures or old flight holes.

The eggs hatch in about four weeks. Larvae, up to 6 mm long, may then burrow through the wood for over two years, leaving in their trail a bore dust (frass) of minute ellipsoidal pellets resembling fine sand.

The larvae finally come to rest in small chambers just below the surface of the timber, where they pupate (change into chrysalides).

Once the transformation to beetle form is complete (between May and August), they bite their way out, leaving holes of about 2 mm in diameter known as 'flight' or 'exit' holes.

After mating, the females will lay their eggs to complete their life cycle.

Because of these beetles' ability to fly, very few timbers are exempt from attack. Some West African hardwoods do, however, seem to be either immune or very resistant to attack – they are:

- Abura
- African walnut
- Afrormosia
- Idigbo

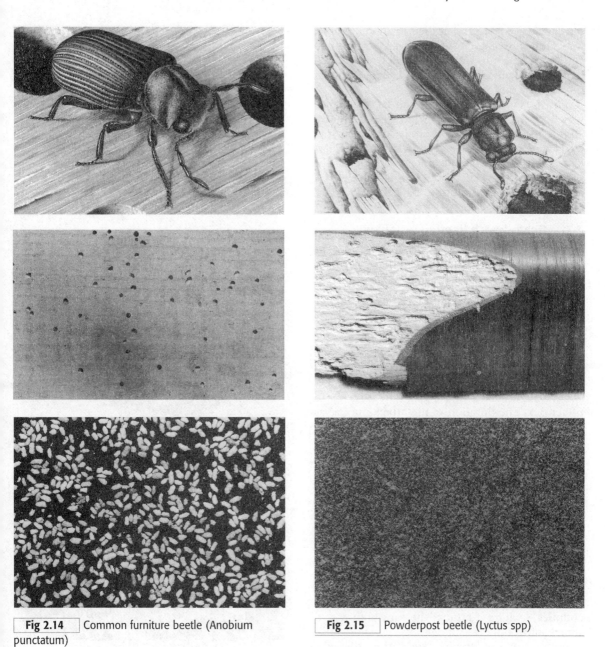

Fig 2.14 | Common furniture beetle (Anobium punctatum)

Fig 2.15 | Powderpost beetle (Lyctus spp)

- Iroko
- Sapele

Evidence of an attack is provided by the flight holes and, in many cases, the bore dust ejected from them. Old, damp, and neglected parts of property, including outbuildings, provide an ideal habitat for these insects – attics and cellars are of particular interest to them, as they may remain damp and undisturbed for many years.

2.3.4 **Powderpost beetle** (*Lyctus spp.* – Fig 2.15)

There are many species of 'Lyctus' – the most common of which is Lyctus *brunneus,* to which this text applies. 'Powder-post' refers to the very fine powdered bore dust (frass) left by the larva of the Lyctus beetle.

This beetle attacks the sapwood of certain hardwoods such as ash, elm, oak, walnut, obeche, and ramin, doing so when they are in a partially dried state and at their most vulnerable, for example, stacked in a timber yard. Infestation can be passed into the home in newly acquired hardwood furniture or recently fixed panelling or block floors and so on.

The life cycle of this insect can be very short, which means that infestation can spread rapidly.

Wood species with small pores (which restrict egg laying) or insufficient starch (essential for larval growth) are safe from attack. Generally, softwoods, some hardwoods with small pores, and heartwood are immune.

2.3.5 Death-watch beetle (*Xestobium rufovillosum* – Fig 2.16)

This is aptly named because of the ticking or tapping noise it makes during its mating season, from March to June, and its often eerie presence in the rafters of ancient buildings and churches.

Its appearance is similar to that of the common furniture beetle, but that is where the similarity ends. Not only is it twice as long, it generally only eats hardwoods which have previously been attacked by fungi – and it seems to prefer oak. The duration of the attack may be as long as ten years or more – much depends on the condition of the wood, the amount of fungal decay, and environmental conditions.

Because the life cycle can be lengthy, infestation over wide areas is limited.

2.3.6 House longhorn beetle (*Hylotrupes bajulus* – Fig 2.17)

This large beetle takes its name from its long feelers. It is a very serious pest on the mainland of Europe and in parts of Britain – mainly in north Surrey and certain areas of adjoining counties – where it attacks the sapwood of seasoned softwood.

Because of the larvae's size and of the subsequent bore holes left by extensive tunnelling over a number of years, the extent of the damage caused to structural timbers has led to much concern in Britain – so much so that it is mandatory, under the Building Regulations, to treat all softwood used for the purpose of constructing a roof or ceiling within those geographical areas stated in the regulations (mainly Surrey and Berkshire) with a wood preservative.

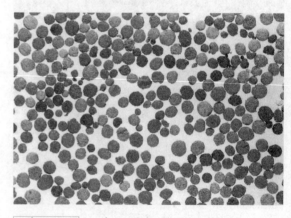

Fig 2.16   Death-watch beetle (Xest obium rufovillosum)

2.3.7 Wood-boring weevils (*Pentarthrum huttoni* and *Euophryum confine* Fig 2.18)

Beetles in this group have distinct protruding snouts from which their feelers project. The two

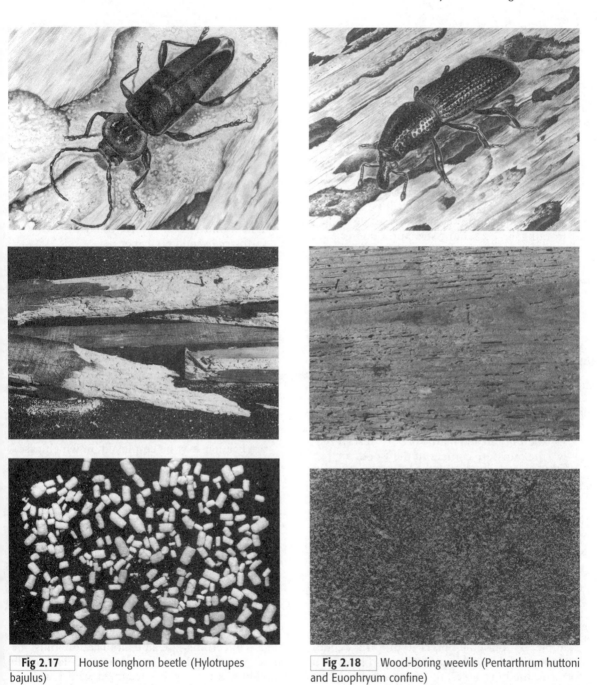

Fig 2.17 House longhorn beetle (Hylotrupes bajulus)

Fig 2.18 Wood-boring weevils (Pentarthrum huttoni and Euophryum confine)

species which concern us here are *Pentarthrum huttoni* and *Euophryum confine* (known also as the *New Zealand weevil*). Their attacks are usually confined to the damp and/or decayed (usually wet rot) sapwood and occasionally heartwood of both softwood and hardwood.

Unlike the other wood boring beetles mentioned, both the adult and larvae of these insects carry out 'boring' activities. Tunnels tend to follow the pattern of the wood's grain.

2.3.8 Prevention and general eradication of wood boring insects:

a **Prevention of attack by wood boring insects** – probably the best preventative measure against insect attack is, where practicable, to keep the m.c. of wood below 10 %, (the interior of most modern centraly heated houses should meet this requirement) for at this level these insects will be discouraged from breeding. As can be seen from figure 1.28, most structural timber members of a building exceed this m.c., therefore if the building has a history of insect attack or is sited in a geographical area where infestation is common (as is the case in certain areas in the United Kingdom) then suitable preservative methods should be considered.

b **General eradication of wood boring insects** – both obvious and suspected areas of activity should be fully investigated to determine:
   i   the extent and nature of the attack,
   ii   the size and shape of flight holes,
   iii the amount and nature of bore dust (frass),
   iv   the moisture content of the wood,
   v   if fungal attack is in evidence,
   vi   the species and nature of the wood under attack,
   vii signs of activity by taking small woods samples.

It should then be possible to establish which beetle or beetles are responsible for the attack.

False alarms can occur if flight holes are the only symptom, as an attack may have taken place before installation, when the timber was 'green' (not dried) – in which case, the culprits would have been killed during timber drying. For example, by 'pin-hole borers' (ambrosia beetle) that take their nourishment from a fungus which forms in the bore holes, this fungus cannot survive in wood with a moisture content lower than about 30% – the beetle and its activity are therefore eliminated once the timber has been dried, but evidence of its attack remains in the shape of very small pin-holes, usually with a dark stain around their perimeter (covering tunnel walls) – also see section 2.3.1.

However, assuming the outbreak being investigated is active, the following measures should be taken:

● Where possible, open up affected areas – cut away wood that is badly attacked, carefully remove it from the site and dispose of it in a safe manner.
● If load-bearing timbers are affected, seek professional advice about their structural stability. Repair or replacement of such items usually involves propping and shoring, to safeguard against structural movement or collapse.
● Thoroughly clear the whole area of debris. Dust removal may involve the use of an industrial vacuum cleaner.
● If the attack is localised to a small area, treat the replacement and adjacent timbers with an approved preservative (insecticide). On the other hand if the attack is more extensive, it would be advisable that the work be undertaken by a firm specialising in insect infestations. In either case, any application of a wood preservative must be carried out according to the manufacturers instructions and the Health and Safety at Work Act guidelines. The method of application will depend on the amount of treatment required. With a small outbreak brush application may be employed, in which case special attention should be given to end grain, fissures, and joints.

Operatives must wear an approved face-mask for protection, to prevent inhaling vapours given off', protective clothing, goggles, and gloves must be used, thus protecting eyes and skin from coming into contact with the materials. Precautions must also be taken against the possibility of a fire, as, at the time of application, organic-solvent types of preservative give off a vapour that is a fire risk.

Preservative manufacturer's recommendations concerning use and application must be strictly adhered to at all times. Habitable dwellings must not be occupied whilst work is undertaken, nor should it be re-occupied until the recommended time span has elapsed, after work is complete (also see section 3.3).

2.3.9 Termites

Termites are a major problem mainly in the tropical and subtropical regions of the world –

they do not appear in the UK. These insects are capable of destroying timber at an alarming rate.

The life-style of termites is somewhat similar to that of ants; for instance, they live in colonies containing workers who carry out the destruction and soldiers who guard and protect the colony from intruders, while the kings and queens are responsible for populating the colony. Similarities, however, end there.

There are two groups of termites:

1 those which dwell entirely inside the timber they are destroying;
2 those which dwell in the ground, in earth cavities, or under mounds of earth, away from their food supply. Access to this food supply is gained by the termites either tunnelling under or over any obstacle they may meet, in order to avoid the light and detection.

a **Prevention of termite attack** – some timbers are resistant to termite attack – for example, Afrormosia, Iroko, and Teak, (there are others) but most timbers can be given good protection by impregnating them with preservative.

Where ground-dwelling termites are a problem, buildings can be designed with physical barriers against access, such as metal or concrete caps between timber and ground, and protruding perimeter sills (shields) of metal or concrete. All building perimeters must be kept clear of shrubbery etc., which might act as a bridge which the termites could use to bypass the perimeter shield.

b **General eradication of termite attack** – methods of eradication will depend on the termite group, i.e. wood or earth dwelling.

With wood-dwelling types, treatment will be similar to that as for other wood-boring insects, but with earth-dwelling types the problem is finding the point of entry into the building. One solution is to trench all round the building and treat the ground with chemicals – the termites still in the building will be cut off from their nest.

3

Wood Preservation and Protection

Except where timber is used in its natural state for practical or economical reasons, it is generally treated with either paint, water-repellent stain, preservative, or special solutions to either reduce or retard the effects of:

a weathering (erosion and discolouration)
b exposure to sunlight
c moisture movement,
d fire;

and to reduce or even eliminate the risk of:

e fungal attack,
f insect attack.

Some of the reasons for using a recognised method of treatment are illustrated in figure 3.1, others include solutions to prevent concrete sticking to surface of timber and wood products (section 3.6.1)

3.1 Paint and varnishes

Exterior-quality paints and varnishes can protect timber from:

● the entry of water,
● abrasive particles (grit and dirt),
● solar radiation – discolouration (not *clear* varnishes).

The weathering effect of wind, rain, and sunlight will gradually degrade the surface of unpainted timber by breaking down its surface. Water penetration of non-durable wood is the greatest problem – not only will it result in variable dimensional change (expansion on wetting, contraction on drying), it also increases the risk of fungal attack. Sunlight causes the

wood to change colour and contributes to its degradation.

Correctly painted surfaces should give relatively good protection for up to five years, provided that:

● the timber is thoroughly dry (i.e. has the appropriate moisture content) before paints are applied,
● the surfaces have been prepared correctly, the paint manufacturer's recommended number of coats are given,

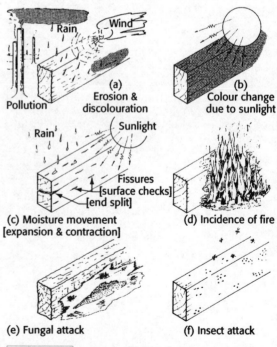

(a) Erosion & discolouration
Pollution

(b) Colour change due to sunlight

(c) Moisture movement [expansion & contraction]
Fissures [surface checks] [end split]

(d) Incidence of fire

(e) Fungal attack

(f) Insect attack

Fig 3.1 Reasons why it is often necessary to protect timber and woodwork

- the end grain has received particular attention with regard to coverage,
- inspections are carried out regularly.

If, as shown in figure 3.2, a painted surface becomes damaged, moisture may enter the wood and become trapped behind the remaining film of paint, and further paint failure and wood decay could quickly follow.

Failure may also be due to joint damage or movement resulting in hairline fractures of non-flexible paint film. This could allow moisture intake by capillary action.

Two ways of reducing the effect of moisture intake are:

1   By using paint which leaves a flexible film over the surface of the wood, then should

any small amount of movement occur this film can stretch as required.

2   As shown in figure 3.3, by using a 'breather' 'or micro-porous' type of paint which allows any trapped moisture, or water vapour which would otherwise become trapped behind the paint film to escape through its surface via minute perforations.

Where micro-porous finishes are used all under-coats (primers and base coats) must also be of the breather type.

Exterior fittings and fixings associated with joinery where micro-porous types of paint have been used, should, because of the possible for-mulation (emulsions), and paint's ability to release water vapour, be rust proofed.

3.2   Water-repellent exterior stains

Exterior stains provide a clear or coloured, matt or semi-gloss, water-repellent surface which – unlike paint – allows the wood grain to show through. Many not only give protection against weathering, but depending on the stain quality, can also protect for up to four years or more against fungal staining due to a wood-preservative content.

One of the main advantages over some paint and varnishes (with the exception of those which are microporous) is that they allow the wood to breathe – thereby allowing trapped moisture to escape, and avoiding the problems shown in fig-ure 3.3(a).

3.3   Preservatives

Unless treated with a preservative, sapwood and not-durable heartwood will be liable to attack by insects and, if the moisture content is above 20%, by fungi.

Other factors may also have to be taken into account, for example:

- Decide on the siting of the timber or woodwork in relation to a hazard situation as shown in table 3.1.
- Determine the natural durability of the wood (section 1.10e) – don't forget all sapwood is classed as 'not durable'
- Check the treatability of the wood (section 1.10f)

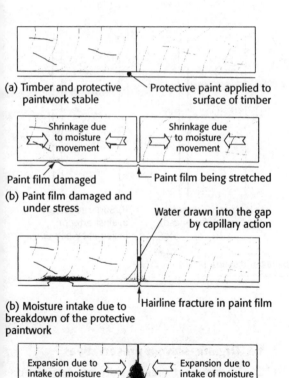

(a) Timber and protective paintwork stable — Protective paint applied to surface of timber

Shrinkage due to moisture movement   Shrinkage due to moisture movement

Paint film damaged — Paint film being stretched
(b) Paint film damaged and under stress

Water drawn into the gap by capillary action

(b) Moisture intake due to breakdown of the protective paintwork — Hairline fracture in paint film

Expansion due to intake of moisture   Expansion due to intake of moisture

(d) Possible result:
- total break-up of protective paint film
- high moisture intake & retension
- timber distortion
- fungal decay

Fig 3.2   How exterior protective paintwork can become non-effective

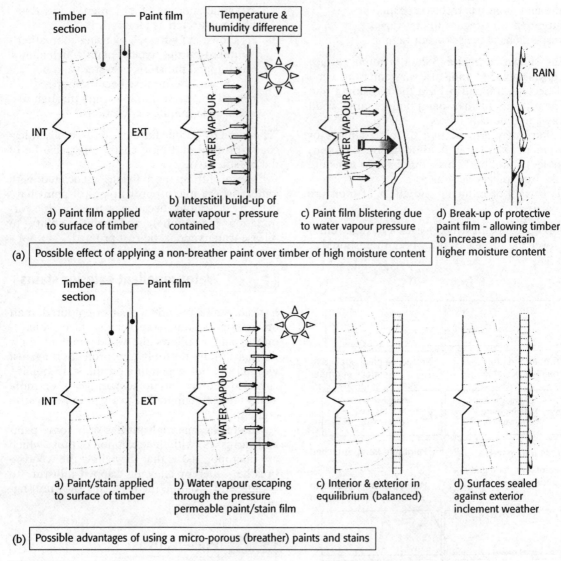

a) Paint film applied to surface of timber

b) Interstitil build-up of water vapour - pressure contained

c) Paint film blistering due to water vapour pressure

d) Break-up of protective paint film - allowing timber to increase and retain higher moisture content

(a) | Possible effect of applying a non-breather paint over timber of high moisture content |

a) Paint/stain applied to surface of timber

b) Water vapour escaping through the pressure permeable paint/stain film

c) Interior & exterior in equilibrium (balanced)

d) Surfaces sealed against exterior inclement weather

(b) | Possible advantages of using a micro-porous (breather) paints and stains |

Fig 3.3   Micro-porous (breather) paints and stains

● Select a suitable preservative to meet the potential hazard class as shown in table 3.1.

Wood preservatives are solutions containing either an insecticide, fungicide, or both. Which are either applied or introduced into the wood to make it toxic (poisonous) to insects and/or fungi.

There are three preservative groups:

i   Organic-solvent types (OS)
ii   Water-borne types (WB)
iii   Tar-oil types (TO).

3.3.1 Organic-solvent preservatives

Use a medium of organic solvents to transmit the toxic chemicals into the wood. After application, the solvents evaporate, leaving the wood toxic to insects and/or fungi.

Examples of its general use are given in table 3.2.

Methods of application include:

● Low pressure processes (double vacuum)
● brushing,

Table 3.1   Hazard classification in relation to siting, exposure and moisture content

Hazard classification	Siting (situation)	Exposure	Moisture content (m.c) of timber
1	Above the ground – covered (protected)	Permanently dry	Permanently below 18% m.c.
2	Above the ground – covered – at risk of becoming wet	Occasionally wet	Occasionally above 20% m.c.
3	Above the ground – uncovered (not protected)	Frequently wet	Frequently above 20% m.c.
4	In contact with the ground and/or fresh water	Permanently wet	Permanently above 20% m.c
5	In contact with salt water	Permanently wet	Permanently above 20% m.c.

Table 3.2   General use of organic solvent (OS) preservative treated timber

Methods of treatment/application	Examples of general use
Low pressure (double vacuum)	**Pitched roofs:** *Wall plates, trussed rafters, ceiling joists, rafters, ridge boards, purlins, binders, sarking (cladding).* **Note:** *Barge boards, fascia boards, and soffits will require a maintained weatherproof protective coating.* **Flat roofs:** *Wall plates, Joists, outriggers, strutting, decking.* **Note:** *Fascia boards and soffits will require a maintained weatherproof protective coating.* **Floors:** *Ground floor joist* **External walls:** *Studs, noggings, head binders, sheathing, external battens and counter battens.* **Note:** *External cladding boards will require a maintained weatherproof protective coating.* **External joinery:** *Doors, windows, porches (excluding sole plates), canopies (inc., roof menbers and trims).* **Note:** *Exposed surfaces will require a maintained protective coating*
Non-pressure: Brushing	**Localised treatment:** *Interior woodwork, or exterior woodwork before the application of an exterior protective finish.* **Remedial work:** *Interior woodwork, or exterior woodwork before the application of an exterior protective finish.*
Spraying	**Remedial work:** *Interior woodwork*
Immersion (dipping)	**Small items of joinery and short lengths of timber**

Note: Preservative formulations will vary to suit the method of application and degree of protection against fungi and/or insect attack.

- spraying,
- immersion.

The solvents used are generally volatile and flammable, and extreme care must therefore be taken at the time of application, and in storing containers.

These preservatives *do not* affect the dimensions of timber (cause swelling) or have a corrosive effect on metals. The ability to glue or paint timber is unaltered after treatment once the preservative has dried.

3.3.2 Water-borne preservatives

Use water to convey the toxic chemicals. There are three types:

1   One for treating unseasoned (green) timber' where the preservative is introduced into the wood by a method known as 'diffusion'.
2   Types requiring a pressure treatment system.
3   Special formulations to allow for in situ and remedial application by non-pressure methods

These preservatives are non-flammable. Drying is always necessary after treatment. When dry treated timber can be painted over.

Examples of its general use are given in table 3.3

3.3.3 Tar-oil preservatives

Derived from coal tar, and are ideal for preserving exterior work which is not to be painted. Examples of its general use are given in table 3.4.

They do not usually have any corrosive effect on metals, but they will stain most porous materials they contact.

The most common form of tar-oil preservative is *creosote*, which is light to dark brown in colour, and can be applied by various processes including brushing and spraying. It has a strong odour for some time after its application.

3.4   Methods of applying preservatives

Preservatives are applied by one of two general methods: pressure or non-pressure. Figure 3.4 gives some indication as to how the method used affects the depth of penetration.

3.4.1 Pressure methods

Timber is put into a sealable chamber into which preservative is introduced under pressure as described below:

High Pressure Methods:

- Full-cell process
- Empty-cell process

Low Pressure Method:

- Double-vacuum process

a   **Full-cell process** – After sealing the chamber, the air is removed by using a vacuum pump. With the chamber still under a vacuum and after a prescribed period, preservative is introduced, filling the chamber. Vacuum is then released and pressure applied. The chamber will remain filled until the timber has absorbed sufficient preservative, which will vary according to the permeability of the wood. Pressure is then released on completion, surplus preservative is pumped out of the chamber back into its storage tank. Finally, a further vacuum is set up in the chamber, but only of sufficient strength to remove any surface preservative – cell cavities will remain full. This process may use creosote or water-borne preservatives.

b   **Empty-cell (Lowry) process** – In this case, the preservative is pumped into the chamber subjected to normal air pressure. After the wood has absorbed sufficient preservative, the pressure is released and surplus preservative is driven out of the wood by the expanding air in the wood cells. A vacuum is then used to draw off any residue. Although the wood cells are emptied, their walls remain fully treated. This process is used for both creosote and water borne preservatives.

c   **Double-vacuum process** – (A typical treatment plant is shown in fig. 3.5). The chamber is sealed and a partial vacuum is created. The chamber is then filled with preservative and pressurised to atmospheric pressure, or above, depending on the process. After a prescribed period, it is then drained and a final vacuum is created to remove excess preservative from the timber. This process is mainly used to apply organic-solvent preservatives into timber for exterior joinery – required depths of penetration are therefore less than for the full-cell and empty-cell processes but better than for immersion. It does not cause timber to swell or distort, and all machining should be done before treatment.

Table 3.3 General use of water-borne (WB) preservative treated timber

Methods of treatment/application	Examples of general use
Low pressure (double vacuum) – example, 'Vacsol Aqua'	**Pitched roofs:** *Wall plates, trussed rafters, ceiling joists, rafters, ridge boards, purlins, binders, sarking (sheet roof covering - prior to weather proofing e.g., tiles / slates etc.).* **Note:** *Barge boards, fascia boards, and soffits will require a maintained weatherproof protective coating.* **Flat roofs:** *Wall plates, Joists, outriggers, strutting, decking.* **Note:** Fascia boards and soffits will also require a maintained weather proof protective coating. **Floors:** *Ground floor joist* **External walls:** *Studs, noggings, head binders, sheathing, external battens and counter battens.* **Note:** *External-cladding boards* will also require a maintained weatherproof protective coating. **External joinery:** *Doors, windows, porches (excluding sole plates), canopies (inc., roof menbers and trims).* **Note:** *Exposed surfaces will require a maintained protective coating*
High pressure (empty cell 'Lowry' process) – for example: 'Tanalised'	**As above,** also, primarily associated with timbers prone to dampness and frequent wetting: *Structural building members* *Agricultural buildings, boundaries and gates* *Horticultural timbers* *Structural items associated with garden features, playground equipment, and landscape design.* *General fencing posts (stakes) and rails.* *Railway sleepers* *Marine work* *Cooling tower timbers*
***Non-pressure:** Brushing	**Localised treatment:** *Interior woodwork, and exterior woodwork before the application of an exterior protective finish.* **Remedial work:** *Interior woodwork, or exterior woodwork before the application of an exterior protective finish.*
Spraying	**Remedial work:** *Interior woodwork*
Immersion (dipping)	**Small items of joinery and short lengths of timber**

*Special formulation

Note: Preservative formulations will vary to suit the method of application and degree of protection against fungi and/insect attack

3.4.2 Non-Pressure process

With the exception of the 'hot and cold open-tank treatment' and 'diffusion' methods described below. Depth of preservative penetration with non-pressure methods is often limited to just below the surface of the wood – as shown in figure 3.4 (a & b).

The following methods of application should be read in conjunction with figure 3.6.

a **Brushing** – Can be used for applying creosote, organic solvent, and some water-borne types of preservative but because of low penetration, it is not a suitable method for timbers that come into contact with the

Table 3.4 General use of tar-oil (TO) type (Creosote) preservative treated timber

Methods of treatment/application	Examples of general use
High pressure – *full cell process*, or *empty cell process* where bleeding of preservative is to be avoided.	**Fencing posts** (stakes), **and rails:** *Farms, estates, domestic gardens, railways and motor-way embankments, etc.* **Poles in the round:** *Transmission and telegraph poles.* **Piling:** *Retaining walls, sea defences, jetties etc.* **Timber sleepers:** *Railways*
Non-pressure: Hot and Cold tank open-tank treatment	**Fencing posts** (stakes), **and rails:** *Farms, estates, domestic gardens, etc.*
Immersion *(steeping)*	**Items above ground:** *Fence panels, rails etc.*
Immersion *(dipping)* The amount of protection offered by all the above treatments will to a greater extent depend on: • the duration of submersion • species of timber • treatability (see Section 1.11(f)) • amount of sapwood / heartwood	**Items above ground:** *Fence panels, rails etc.*
Brushing	**Remedial work:** *Fence panels, rails etc.*

Note: Preservative formulations will vary to suit the method of application and degree of protection against fungi and/insect attack.

ground. As a rule, re-treatment is advisable every three to four years.

b **Spraying** – Similar penetration and conditions apply as for brushing. Because of the health risk associated with applying preservatives, precautions should always be taken – particularly when spraying – to ensure that:

• only coarse sprays are used, to avoid atomisation;
• work areas are well ventilated;
• operatives are suitably clothed;
• hands are protected by gloves;
• mouth and nose are protected by an approved face-mask;
• eyes are protected with snug-fitting goggles – not glasses;
• Manufacturer's instructions are followed.

c **Deluging** – In deluging, the timber is passed through a tunnel of jets that spray it with preservative.

d **Immersion**

• *Dipping* – the timber is submerged in a tank of preservative (coal-tar oils or organic-solvent types) for a short period, then allowed to drain.
• *Steeping* – The timber is submerged for periods ranging from a few hours to weeks, depending on the wood species, the sectional size of the timber, and its end use. Steeping is a suitable method for preserving fence posts etc

e **Hot & Cold (open tank treatment)** – The timber is submerged in a tank of preservative (coal-tar oil) which is heated. It is then allowed to cool in the tank or is transferred to a tank of cold preservative. This treatment is suitable only for permeable timbers and sapwood. Coal-tar oils are flammable, therefore extra care is necessary with regard to the heat source.

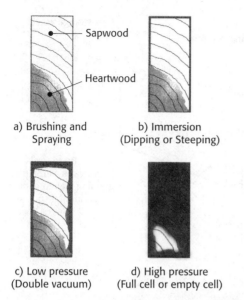

a) Brushing and   b) Immersion
   Spraying   (Dipping or Steeping)

c) Low pressure   d) High pressure
   (Double vacuum)   (Full cell or empty cell)

Fig 3.4   A guide to preservative impregnation in relation to different methods of application

f **Diffusion** – This method of treatment is only associated with freshly felled green timber which, at the saw mills, is immersed in a water-borne preservative (usually boron salts) and then close-piled and placed under cover until the preservative has diffused into the wood. The type of water-borne preservative used in this process is liable to leach out from the wood, which makes the timber unsuitable in wet locations unless an impervious surface treatment is given, i.e. paint or varnish.

3.5   Flame-retardent treatments

By impregnating the timber with a solution of various salts, other chemicals, or coating its surface with special paints, it is possible to reduce the rate at which a flame would normally spread over its surface.

Timber impregnated with flame-retarding salts is not normally suited to exterior use, because the salts are liable to leach out. Strength properties of timber will be reduced as a result of this treatment.

By using a special process, leaching can be avoided and strength properties of the timber remain unaltered.

3.5.1 Paint & varnishes

Some paints and varnishes are 'intumescent' – they swell when subjected to heat and protect the wood by forming an insulating layer over the surface of the timber. Others give off a gas which protects against flaming

3.6   Other treatments

Other treatments applied to timber include mould oils and release agents.

3.6.1 Mould oils and release agents

These solutions are applied to the inner surfaces of formwork (parts which will come into contact with wet concrete) to prevent concrete sticking to them. A released agent (sometimes called a parting agent) may be in the form of an oil (mould oil) emulsion, a synthetic resin or plastic compound to give those surfaces a hard, abrasive resistant protective coating, which may also require mould oil treatment.

Treatment as a whole increases the life of the formwork and helps reduced the number of surface blemishes and 'blow-holes' (holes left by pockets of air) appearing on the surface of the concrete.

Some treatments are not suitable for metal formwork, as they tend to encourage rusting. The manufacturer's recommendations should always be observed concerning treatment use and application. Method of application may include spraying, but mould oils are more usually applied by swab or brush.

Note: care should be taken not to get mould oil (release agents) on any steel reinforcement or areas where adhesion is important.

3.7   Health & safety

With all chemical products, care must be taken by operatives during application, and when handling treated products. Each product should be dealt with according to the manufacturers or suppliers' instructions and safety guidelines issued with each product.

At all times the requirements of the Health and Safety at work Act and the Control of Substances Hazardous to Health regulations

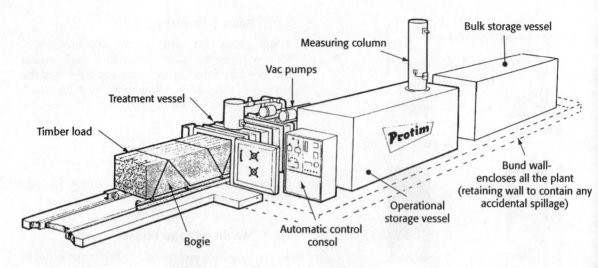

Bulk storage vessel

Measuring column

Vac pumps

Treatment vessel

Timber load

Protim

Bund wall-
encloses all the plant
(retaining wall to contain any
accidental spillage)

Operational
storage vessel

Automatic control
consol

Bogie

Fig 3.5   'Protim' double vacuum treatment plant (courtesy of Protim Ltd)

(COSHH) must be implemented. Particular attention should be given to the application of insecticides, fungicides and flame retardant chemicals and products treated with them.

3.7.1 Remedial treatment

Listed below are some general safety guidelines when using wood preservatives particularly during remedial (repair) work:

1 Protective clothing must be worn, to include a full overall (coveralls) and protective PVC/Synthetic rubber gauntlets.
2 Protective footwear must be resistant to the chemicals used, and with non-slip soles.
3 A facemask (of the appropriate approved type) must be worn when cleaning prior to treatment and for general application of a preservative.
4 A respirator (of the approved type) must be

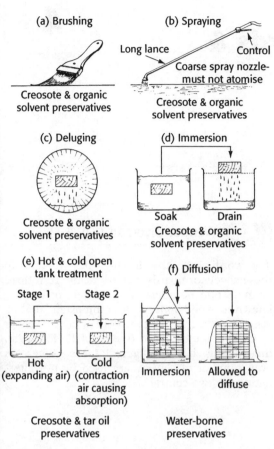

(a) Brushing

Creosote & organic
solvent preservatives

(b) Spraying

Long lance   Control
Coarse spray nozzle-
must not atomise

Creosote & organic
solvent preservatives

(c) Deluging

Creosote & organic
solvent preservatives

(d) Immersion

Soak   Drain
Creosote & organic
solvent preservatives

(e) Hot & cold open
tank treatment

Stage 1   Stage 2

Hot   Cold
(expanding air) (contraction
           air causing
           absorption)

Creosote & tar oil
preservatives

(f) Diffusion

Immersion   Allowed to
             diffuse

Water-borne
preservatives

Fig 3.6   Illustrative guide to different non-pressure methods of applying wood preservatives

used when spraying insecticide and or fungicide solutions.

5 Goggles or a full-face visor must be worn during the application of preservatives.

6 All practical means should be taken to ensure good ventilation when applying preservatives.

7 In the event of the preservative coming in contact with the skin, it must be washed off immediately. Should preservatives come in contact with the eyes wash them immediately with clean fresh water for 10 to 15 minutes and seek medical advice.

8 Have a normal bath or shower of the whole body after completion of the work.

9 Precaution should be taken to prevent people and/or animals from gaining access to areas contaminated by preservatives.

10 Drinking or eating must be prohibited whilst applying preservatives or working with treated material.

11 Hands and exposed skin must be thoroughly washed after work and before eating, smoking, or going to the toilet.

12 Smoking is strictly prohibited in areas being treated, or where treatment has been recently carried out, or whilst handling treated material.

13 Newly treated areas must not be reoccupied until the specified period time given by the manufacture has elapsed.

14 Do not contaminate water tanks, the ground, or water courses (dangerous to all aquatic life),

15 Solutions and their containers must be disposed of in a safe way.

16 Avoid all contact with plant life.

17 Do not apply to surfaces where food is stored, prepared or eaten.

18 Keep unused preservatives in their original container – sealed and in a safe place.

19 Before treating any structure used by bats, English Nature Must be notified – all bats are protected under the 'Wildlife & Countryside Act 1981'.

Note: extensive remedial work should only be carried out by firms that specialise in this type of work, and who have employees who are specially trained and fully equipped to cope with all eventualities.

Manufactured Boards & Panel Products

4

We have seen how wood can be subject to dimensional change and distortion when used in its solid state, particularly as a wide board. It is this inherent problem, together with its cost, that often restricts the use of wood where wide or large areas have to be covered. This is the kind of work where manufactured boards are mainly used.

For the purpose of this chapter, 'manufactured' boards will be taken to mean those boards and sheet materials, the greater part of which may be composed of wood veneer, strips, particles, or their combination, such as:

- **Plywoods:**
  - Veneer plywood,
  - Core plywood-laminated boards (laminboard, blockboard, and battenboard).
- **Particle boards:**
  - Wood chipboards,
  - Flaxboard
  - Bagasse board
  - Cement bonded particle board.
- **OSB** (orientated strand board).
- **Fibre boards:**
  - Softboard,
  - Medium board,
  - Hardboard (standard & tempered)
- **Medium Density Fibreboard** (MDF),

Or other board and sheet materials often used by joiners as part of the job. For example:

- **Laminated plastics**
- **Fibre cement building boards**
- **Plasterboards**
- **Composite boards**

4.1 Veneer plywood (Fig. 4.1)

The word plywood is usually taken to refer to those sheets or boards that are made from three or more odd numbers of thin layers of wood – known as wood veneers. It is important that all veneers on each side of the core or centre veneer or veneers are balanced (see fig. 4.1). However, there are, as shown in figure 4.2, cases where an even number of veneers are used by not alternating the two central veneers.

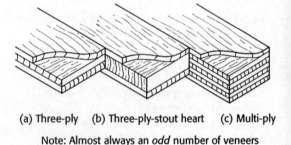

(a) Three-ply   (b) Three-ply-stout heart   (c) Multi-ply

Note: Almost always an *odd* number of veneers

| Fig 4.1 | Veneer plywood |

Neutral axis — Odd–Number of veneers   Glue line — Even–Number of veneers (plies 6) Odd–number of layers

| Fig 4.2 | Number of veneers in relation to the number of plywood layers |

4.1.7 Plywood Board markings (Fig 4.4)

Each board (sheet) of plywood should be marked on its face with several identification details such as:

- The manufacturers name, or logo (trademark).
- Compliance standard, for eg, BS EN Number
- Type of board
- Nominal thickness
- Formaldehyde release class (formaldehyde emissions – 'A' = lowest 'C' = highest)
- Quality stamp/label – any certification body.

4.1.8 Check list for when ordering veneer plywood:

- Grade of glue (adhesive) bond class 1–3 (Interior or Exterior)
- Face veneer (type) and orientation (see fig. 4.7)
- Board (sheet) *size – your first dimension should indicate the direction of the face grain as shown in figure 4.7. In figure 4.7(a) cross grain is stated first, and in figure 4.7(b) long grain is stated first.
- Nominal (manufactured size – not necessarily the finished size) thickness
- Number of veneers (three-ply or multi-ply)

- Surface finish and condition (section 4.1.3)
- Availability

*A range of available sizes is given in appendix 2.

4.2   Core plywood

Plywood consisting of wood strips contained by two or more wood veneers on both sides. It is the width of these strips that gives the board the name of:

- Laminboard, or
- Blockboard

4.2.1 Laminboard (Fig 4.8)

Has a core made up of a lamination of narrow wood strips (veneer on edge) – not exceeding 7 mm wide – glued together, then faced with one or two wood veneers on each face.

The methods of laying up using 5 ply veneer construction are shown in figure 4.8.a–c.

a Both side veneers are at right angles to the core.
b Inner veneers are at right angles to the core – cross-banded outer facing veneers run in the same direction of the core grain.
c As above, except that the facing veneers run in the opposite direction of the core grain.

Laminboard has a virtually distortion (ripple)-free surface.

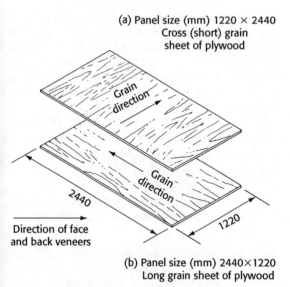

(a) Panel size (mm) 1220 × 2440
Cross (short) grain
sheet of plywood

Grain direction

Grain direction

2440

1220

Direction of face and back veneers

(b) Panel size (mm) 2440×1220
Long grain sheet of plywood

| **Fig 4.7** | Direction of grain in relation to a stated sheet size |

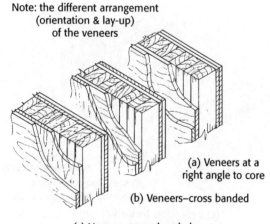

Note: the different arrangement
(orientation & lay-up)
of the veneers

(a) Veneers at a right angle to core

(b) Veneers–cross banded

(c) Veneers–cross banded

| **Fig 4.8** | Laminboard, 5-ply construction |

4.2.2 Blockboard (Fig. 4.9)

Similar to laminboard, except that the wood strips used in its core are wider – usually between 19 mm and 30 mm.

The methods of laying up using 3 and 5 ply veneer construction are shown in figure 4.9.a–c.

a 3 ply – single outer veneer at right angles to the core. The facing veneers of these boards are not generally suitable for self-finishes, as the wider core can distort slightly and induce a rippled effect on the finished surfaces of the board. They can be used as a baseboard for further facings such as a plastic laminate.

b 5 ply – a more stable board, both outer veneers are cross-banded. Facing veneers run at right angles to the core material. Generally suitable for self-finishes.

c 5 ply – as above , both outer veneers are cross-banded. In this case, the facing veneers run in the same direction as the core material. Generally suitable for self-finishes.

4.2.3 Manufacturing process

The method is similar to that of veneer plywood – wood veneer production being identical to that shown in figure 4.3. The main difference being the laying-up of the core. For example;

● Core material – obtained from saw mill short ends and residue, dried and cut into strips

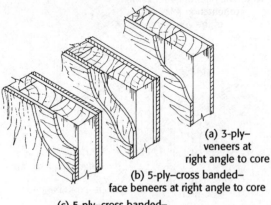

(a) 3-ply–
veneers at
right angle to core

(b) 5-ply–cross banded–
face beneers at right angle to core

(c) 5-ply–cross banded–
face veneers parallel to core

| **Fig 4.9** | Blockboard, 3 and 5-ply construction |

for blockboard. Alternatively, cut from sheets of veneer, veneer plywood for laminboard.

● Core slab – strips are assembled into a slab, they may, depending on board type be pre-glued together.

● Laying-up – veneers are now bonded to the slab usually with a urea formaldehyde adhesive (see table 4.1)

● Pressing and curing of the adhesive under controlled conditions

● Trimming to size, and sanding to the required finish.

● Labelling and dispatching.

4.2.4 End use

These can be classified as utility boards and restricted to those areas that are not to be subjected to dampness. Suitable for interior use and situation shown in table 4.2(b).

Note: It is probably worth mentioning that some years ago there was another core plywood used in the trade called 'Battenboard' the significance here was that the core material was much wider. It therefore follows that this type of board was less stable than blockboard, and subject to more surface irregularities.

4.3 Chipboard

The main natural ingredients that go to make this 'particleboard' are wood particles derived from softwoods such as spruces, pines, firs, and hardwoods such as birch. These particles are mixed with a synthetic resin that binds them together to form boards of various types and thicknesses from 5 to 30 mm.

Wood required to produce these chippings comes from many different sources:

● Young immature trees (often specifically grown as chipping material).
● Forest thinnings.
● Slabs and residue from sawmills (fig. 1.16).
● Wood-machining waste (shavings, chippings, etc).

Other raw materials include the shives (fibrous slivers) from the flax plant to produce a board known as 'Flaxboard'. The fibrous nature of the sugar cane can also be utilised after the sugar content is removed to produce a particleboard known as 'Bagasse board'.

4.3.1 Manufacturing process

Can differ with the type of board and raw material used. However, in simple terms as shown in figure 4.10, the process can involve:

1 Selection of the raw material.
2 Chipping the raw material – surface and core chips may be prepared separately.
3 Drying the woody particles to about 2.5% mc in special dryers to help with gluing and pressing.
4 Grading particles into sizes to suit board type and layers.
5 Blending with a suitable synthetic-resin adhesive (binder) usually urea formaldehyde or MUF (see table 4.1);
6 Board forming – with the exception of 'Extruded boards', pre-coated particles are laid up into mats. Distribution of particles will depend on the board type – mats may be pre-compressed before final pressing.
7 Pressing – mats are either: a) pressed flat between platens of a multi layer hot press, or, b) by a continuous press to achieve the required thickness.
8 Trimming – after cooling, boards are trimmed to size, and allowed to mature, then finally finished by sanding.

9 Sanding – boards are sanded to size.
10 Quality inspection
11 Packaging – stamping, labelling, packaging by various means ready for dispatch.

4.3.2 Type of board construction

There are several types of these pressed boards, such as:

● single-layer construction,
● three-layer construction,
● multi-layer construction,
● graded-density.

a **Single-layer boards** (fig. 4.11(a)) – a uniform mass of particles of either wood, flax, or both. The type, grade, and compaction of these particles will affect the board's strength and working properties – boards may for instance be classed as interior, structural, depending on their use.
    Because of their composition (uniformity of particles), single-layer boards present very few problems when being cut.
b **Three-layer boards** (fig. 4.11(b)) – these consist of a low-density core of large particles, sandwiched between two relatively higher-density layers of fine particles. These boards have a very smooth even surface suitable for direct painting etc. They may be classed as general-purpose or as interior non-structural.
c **Multi-layer boards** (not illustrated) – similar to three-layer but for an increase in

1. Raw material   2. Chip production   3. Drying process

5. Blending with adhesive   4. Grading particles

6. Board forming

7(b). Continuous press

7(a). Pressing multi-layer press

8. Trimming

9. Sanding

10. Quality inspection   11. Packaging/sorting

Fig 4.10   Guide to the processes involved in the manufacture of chipboard

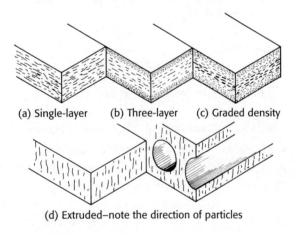

(a) Single-layer   (b) Three-layer   (c) Graded density

(d) Extruded–note the direction of particles

Fig 4.11   Some types of chipboard (particle board) construction

the number of layers, and possibly a layer of higher density core to improve strength.

Care should be taken when cutting these boards, as, unlike single-layer types, they are inclined to split or chip away at the cut edge.

d **Graded-density boards** (fig. 4.11(c)) – these have a board structure midway between the single-layer and three-layer types. Their particles vary in size, getting smaller from the centre outwards. They are suitable for non-structural use and for furniture production.

e **Extruded boards** (fig. 4.11(d)) – the prepared mixture of shredded chippings and adhesive passes through a die, resulting in an extruded board of predetermined thickness and width but of unlimited length. The holes in these boards are made by metal heating tubes, which assist in curing the adhesive, thus enabling much thicker boards to be produced. These holes also reduce the overall weight of the board.

Chipboard produced in this way will have some of its particles located at right angles to the face of the board, thus reducing its strength. However, the main use of these boards is as comparatively lightweight core material to be sandwiched between suitable layers of veneers or other materials to give the required stability.

4.3.3 Grades in relation to end use

Although chipboard can generally be used in situations similar to plywood, its grade and end use will be determined by:

- its method of manufacture,
- the bonding agent,
- special treatment – surface treament (e.g. veneers of wood, melamine, or plastic laminates), or integral (e.g. fire retardants).

Table 4.4 classifies board types (grades) together with possible end use, based on their mechanical performance level and resistance to moisture.

4.3.4 **Chipboard markings** (Fig 4.12)

Each board (sheet) of chipboard should have several identification details marked on its face with indelible printing or label such as:

- The manufacturers name, or logo (trademark).
- Compliance standard, for example, BS EN Number

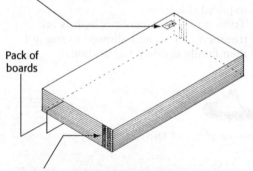

Each board or pack to be marked with at least the
- following information:
- Relevant part BS EN Number
- Manufacturers name, trade mark or identification
- Board type (classification)
- Nominal thickness
- Formaldehyde release class
- Batch number and/or date
Either indelible printing or label may show this information

Pack of boards

Additional edge mark identification via colour stripes–at least 25mm apart towards the corner may be used (see table 4.5)

Fig 4.12   Chipboard labelling and marking (board markings)

Table 4.4   Classification of chipboard types in relation to their use

Classification	Use and possible restrictions
P2 Boards	General purpose boards – only to be used in dry conditions
P3 Boards	Boards suitable for interior fitments – only to be used in dry conditions
P4 Boards	Load bearing boards – only to be used in dry conditions
P5 Boards	Load bearing boards – can be used in humid conditions (example interior floor decking)
P6 Boards	Heavy duty load bearing boards – only to be used in dry conditions
P7 Boards	Heavy duty load bearing boards – can be used in humid conditions

- Type of board – (classification)
- Nominal thickness
- Formaldehyde release class (formaldehyde emissions – 'A' = lowest 'C' = highest)
- Batch number – production week/year.

In some instances, as can be seen from table 4.5., and figure 4.12, board suitability can be recognised by two coloured stripes set 25 mm apart marked vertically towards the corner of the pack. They indicate the boards' suitability for its application and, an acceptable location in relation to the environment that surrounds it.

4.3.5 Check list for when ordering chipboard:

- Board type in relation to its make-up (single layer etc)
- Type classification P2–P7, in relation to end use (Hazard class 1 – 2)
- Board (sheet) *size

- Nominal (manufactured size – not necessarily the finished size) thickness
- Surface finish and condition

*A range of available sizes is given in appendix 2.

4.4 Cement bonded particleboard

A mixture of wood particles (as low as 20%) bonded with Magnesite or Portland cement, to produce two types of high-density board. Used where fire and weather resistances are required, examples of type and use are given in table 4.6.

4.4.1 Manufacturing process

Apart from using cement as a binder, manufacturing processes differ from chipboard in that the logs used as raw material are totally debarked and stored for a period of time before being chipped to neutralise any extractive (see section 1.10.3g)

Table 4.5   Voluntary colour coded stripes for chipboards to indicate suitable use and situation

	Colour of first *stripe	Application (use)
First stripe	White	General purpose boards
	Yellow	Load-bearing boards

	Colour of second *stripe	Environmental condition (situation)
Second stripe	Blue	Dry condition
	Green	Humid condition

* First and second stripe applied vertically 25 mm apart (see fig 4.10)

Table 4.6   Cement-bonded particleboord type (grade) in relation to possible end use

Type of board (grade)	Bonding agent (binder) – Cement	Important properties features	Typical end use
T1		Internal use only	Internal linings, walls etc.
		Low to moderate resistance to moisture	Window boards, casings etc.
		High resistance to fire	
		Liable to attack by wet rot	
		Resistant to insect attack	
T2		Internal and external use	Timber framed buildings (sheathing)
		Good resistance to moisture	External cladding, flooring
		High resistance to fire	Roof decking
		Not liable to attack by wet rot	
		Resistant to insect attack	

that may be present in the wood, which could affect the curing (setting) of the cement binder. The production sequence is similar for example:

- Assembling raw material
- Chip production
- Grading particles
- Blending with cement, water, and any additives
- Board forming
- Pressing and curing
- Trimming to size
- Quality inspection (graded)
- Left to cure (harden) for up to 18 days
- Boards are conditioned to allow their moisture content to become in balance (equilibrium) with the environment that surrounds them.
- Packaged ready for dispatch.

Boards can be satisfactorily cut using both hand and power tools and (with facilities for dust extraction – see section 4.14) with tungsten carbide tipped blades.

4.4.2 Check list for ordering cement bonded particleboard:

- Board type in relation to its end use (interior/exterior) and fixing location (hazard class)
- Availability
- Board (sheet) *size
- Nominal (manufactured size – not necessarily the finished size) thickness
- Surface finish and condition
- Weight consideration with regards to handling and fixing

*A range of available sizes is given in appendix 2.

4.5 Oriented strand board OSB

Oriented strand board (OSB) is a multi-layer (usually three layer) board, each layer being cross-banded in a similar fashion to 3ply-veneer plywood.

Each OSB layer is, as shown in figure 4.13, built up of large thin softwood strands and a binder. Strands in the outer layers are more or less in line with the boards length. Strands in the core layer may be randomly orientated, or roughly in line with the board's width at right angles to the outer layers.

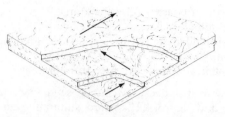

Wood strands lay predominantly in the direction of the arrows

| **Fig 4.13** | Guide to OSB strand orientation |

4.5.1 Manufacturing process

The following sequence as shown in figure 4.14 shows in basic terms how OSB can be manufactured:

1 Softwood logs are sorted as required
2 Debarked and washed
3 Logs are cut to length, they are then waferised into flakes by cutting strands of wood off the log.
4 Flakes are dried in a tumbler drier prior to storage
5 Flakes are blended (coated) with a synthetic resin adhesive (either phenol formaldehyde (PF) or melamine urea formaldehyde (MUF) depending on the board type) and a proportion of wax within a rotary blender.
6 Flakes are arranged into three layered mat (as shown in fig. 4.13) onto a continuous belt – a process known as mat-forming
7 The composition is then compressed to the required density under heat to activate the adhesive
8 After cooling, the boards are trimmed to size then conditioned as required.
9 The finishing line may or may not involve sanding
10 Boards will be marked or labelled according to type, before being packaged ready for dispatch.

4.5.2 OSB type, classification and possible end use

As shown in table 4.7 boards will be marked according to their suitability for use.

Restrictions on use will largely depend on the

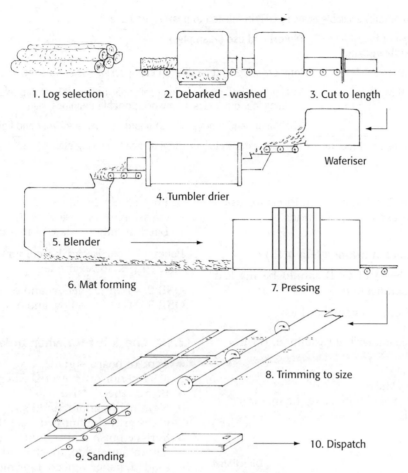

1. Log selection   2. Debarked - washed   3. Cut to length

Waferiser

4. Tumbler drier

5. Blender

6. Mat forming   7. Pressing

8. Trimming to size

9. Sanding   10. Dispatch

Fig 4.14   Guide to the processes involved in the manufacture of OSB

Table 4.7   OSB Types to European Standards

Board mark	Definition
OSB 1	General-purpose board designed for interior use in dry conditions – suitable for interior furniture and fitments, etc.
OSB 2	Load-bearing hoards can only be used in dry conditions
OSB 3	Load-bearing hoards can he used in humid conditions
OSB 4	Heavy duty load-bearing boards can he used in humid conditions

site the board is to occupy. In other words will the board to subjected to conditions which might affect is physical properties? For example, loss of strength, deterioration in its make-up (composition) and so on.

As you will see from table 4.8, the common types OSB 2 & 3 are reliant on situations that keep within condition restrictions set down as:

● Hazard class 1. – Dry conditions with no risk of wetting (generally interior applications).

Table 4.8   Commonly available types of OSB in relation to possible end use

Hazard Class	Type of board – suitable grade	Typical end use (examples)
1	OSB 2	Temporary works (boarding up etc), packing cases, shelving, interior projects etc.
2	OSB 3	Sarking* pitched roofs, decking flat roofs, flooring, sheathing (walls), Site hoarding, cladding, packing cases, formwork, portable buildings, signs
		NB: Tongue and groove (T & G) boards available for roof and floor decking

Note: * Exterior covering of roof rafters provides extra stability and resistance to racking etc.

● Hazard class 2. – Humid conditions for use in protected exterior applications

4.5.3 OSB board markings (Fig 4.15)

Each board (sheet) of OSB should be marked with several identification details such as:

● The manufacturers name, or logo (trademark).
● Compliance standard, for example, BS EN Number
● Type of board
● Nominal thickness
● Directional arrow to indicate the major strength axis

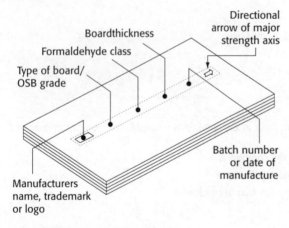

Boardthickness
Formaldehyde class
Type of board/ OSB grade
Directional arrow of major strength axis
Batch number or date of manufacture
Manufacturers name, trademark or logo

Note: Additional information (markings) on tounged and grooved (T&G) boards:
● THIS SIDE UP
● MINIMUM PERIMETER GAP (mm)
● MAXIMUM SPAN (mm)
● MAXIMUM NAILING CENTRES (mm)

Fig 4.15   OSB Labelling and surface marking

● Formaldehyde release class (formaldehyde emissions – 'A' = lowest 'C' = highest)
● Batch number – date of manufacture

Packs of boards can be colour coded by using edge stripes, for example:

OSB 2 – Two (2) yellow and one (1) blue stripe
OSB 3 – Two (2) yellow and one (1) green stripe

4.5.4 Check list for when ordering OSB:

● Type of board – grade to suit requirements ie. (Interior or protected exterior)
● Board (sheet) *size
● Nominal (manufactured size – not necessarily the finished size) thickness
● Surface finish and condition, for example: natural unsanded, touch sanded, fully sanded, paper veneered, or bitumen-coated.
● Availability

*A range of available sizes is given in appendix 2.
   Note: *OSB should not be confused with 'Waferboard'*, waferboard was a flake board made up mainly of hardwood flakes and a binder, all of which were randomly arranged in a single layer and glued together.

4.6   Fibre building boards

Fibre building boards are produced from wood that has been shredded into a fibrous state then reassembled into a uniform sheet form (board). The process used to produce these boards is done by either using a 'Wet' or 'Dry' process. The wet process uses the natural inherent adhesive properties of wood to fuse shredded wood fibres together without the use of any synthetic adhesive. Boards produced using the wet process include

Table 4.13 MDF – in relation to possible end use

Example types of MDF	Approx Density kg/m³	Bonding agent (binder)	General Properties	Typical examples of possible end use
MDF. (Standard)	700	(UF) resin	*Interior use only* Both faces sanded smooth, surfaces suitable for painting, veneering etc.	Furniture and cabinet work, moulded sections, door skins and panels
MDF. (High density)	960			As above & Industrial shelving, flooring systems
MDF. (Flame retardant)	750	polymerised resin plus flame retardant additives	*Interior use only* Fire resistant properties	Wall linings, partitions (non-load bearing), ceilings etc
MDF.H1 (Moisture Resistant)	750	(MUF) resin	*Interior use only:*	Joinery sections – staircases treads and risers, window boards, etc. Moulded sections – skirting boards, architrave's, cornices, etc.;
MDF.H2 (Exterior)	740	Exterior grade polymerised resin	*Interior and *exterior use (when all faces & edges are sealed with an exterior grade of coating)*	Exterior signs and notice boards. Fascia and soffit boards; Shopfronts -exterior moldings, Door parts (raised and fielded panels) etc.

*Moisture resistant for exterior conditions – suitable for use in hazard class 1, 2, & 3 environments.

Table 4.14 Voluntary colour coded stripes for fibreboard's to indicate suitable use and situation

	Colour of first stripe	Application (use)
First stripe	White	General purpose boards
	Yellow	Load-bearing boards

	Colour of second stripe	Environmental condition (situation)
Second stripe	Blue	Dry condition
	Green	Humid condition
	Brown	Exterior conditions

pre-formed (patterned – pierced, embossed etc.)
● Availability

*A range of available sizes is given in appendix 2.

4.7 Laminated plastics (decorative laminates)

Laminated plastics are thin synthetic (man-made) plastic veneers capable of providing both a decorative and hygienic finish to most horizontal and vertical surfaces.

Figure 4.17 shows how the laminations are built-up before being bonded together by a combination of heat and pressure. Final thickness will depend on type of application, for example, post-forming (section 4.7.4) veneers will be thinner than those applied in situ – but not usually thicker than 1.5 mm. Table 4.15 lists some types with possible application.

On work surfaces It is good practice to use laminates which offer some scratch resistance,

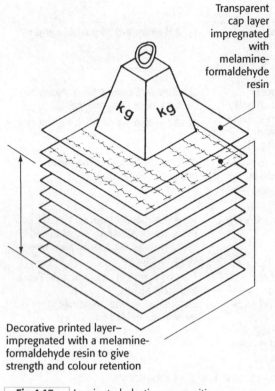

Transparent cap layer impregnated with melamine-formaldehyde resin

Decorative printed layer–impregnated with a melamine-formaldehyde resin to give strength and colour retention

Fig 4.17   Laminated plastics composition

such as those with patterned and light coloured matt and textured surfaces. In contrast, shiny (glossy) and plain dark colours are usually only suitable for vertical application.

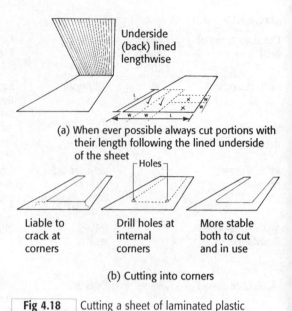

Underside (back) lined lengthwise

(a) When ever possible always cut portions with their length following the lined underside of the sheet

Holes

Liable to crack at corners

Drill holes at internal corners

More stable both to cut and in use

(b) Cutting into corners

Fig 4.18   Cutting a sheet of laminated plastic

4.7.1 Cutting, boring and trimming

These veneers can be successfully cut from the sheet by using hand and/or machine tools. But because of their thin brittle nature they must always be fully supported on either side of each cut, figure 4.18 shows how pieces are cut from a sheet and how cuts are made into corners. Internal corner cuts should be made towards a rounded (radiused) edge, to help avoid corner cracks. More importantly, internal corner stress

Table 4.15   Types of laminated plastics against possible end use

Laminate type	Examples of application
Horizontal general purpose standard grade (HGS)	Working surfaces – kitchens, restaurant tables, counters, and vertical surfaces subject to heavy wear
Vertical general purpose standard grade (VGS)	Wall and door panels, occasional use shelving, display cabinets' etc.
Horizontal general purpose Post forming grade (HGP)	Heavy duty surfaces parts of which are to be curved by post forming – Working surfaces – kitchens, restaurant tables, counters,
Horizontal general purpose flame retardant grade (HGF)	Surfaces which have to meet British Standard requirements for class '1' spread of flame, and/or class' 0' statutory requirements of the Building Regulations when suitably bonded to a suitable substrate
Vertical general purpose flame retardant grade (VGF)	Surfaces which have to meet British Standard requirements for class '1' spread of flame, and/or class' 0' statutory requirements of the Building Regulations when suitably bonded to a suitable substrate
Flame retardant grade for post forming (FRP)	As above when post forming is required

cracks can result from restrained differential movement when the laminate is finally in place in high-risk areas with varying degrees of humidity, such as, an intermittently centrally heated environment. For the same reason, restraining or fixing screws should be used only via movement plates (see sections 12.4), or over-sized clearance holes, as shown in figure 4.19.

Hand Tools – veneers can be cut by using a sharp fine toothed tenon saw, cutting from the decorative face. Or alternatively, by scoring through the decorative face with a purpose made scoring tool, then gently lifting the waste or off-cut side, thus closing the 'V' and allowing the sheet to break along the scored line.

A block plane (low angled blade type section 5.4.2) and/or file can be used to trim edges.

A special note of caution is needed here! Always remember to keep your hands and fingers away and clear of edges whilst this process is being carried out: processed edges can be very sharp!

Machine Tools – Because of its hard brittle nature, special care must be taken with both methods of holding the laminated plastic whilst it is being cut, and during the machining process. Special blades and cutters (tungsten carbide tipped TCT) are available; both the manufacturers of laminated plastics and many toolmakers offer recommendations about blade and cutter peripheral cutting speeds and techniques.

Whilst processing operations are being carried out there is always the risk of injury to the eyes. It is therefore a requirement and essential that eye protec-tion is worn at all times, not only by the operator, but also by others in close proximity to the operation.

4.7.2 Veneer application

Laminated plastics are used as veneers, so their application onto a suitable support (core material) can be dealt with in a similar manner. Almost any board material can be used for this purpose.

If the support material is to retain its shape (i.e. flatness), it must, as shown in figure 4.20, be kept in balance. Therefore any veneer or additional veneers (in the case of plywood) applied to one face (fig. 4.20a) should have an equivalent compensating veneer applied to the opposite face (fig. 4.20b). With laminated plastics, for the best results the balancer (counter veneer) is of the same grade and colour as the face laminate. However, standard balancer veneers to a lower standard are available, and where flatness is not essential, a universal balancer or any backing laminate may be used. In certain circumstances, where an extra thick supported backing is used, as shown in figure 4.20c, a counter veneer may not be necessary.

Unsuitable backings (core materials) include; plastered and cement-rendered surfaces, gypsum plaster board (see section 4.9) and solid timber. Narrow (not exceeding 75 mm wide) quarter sawn boards could be acceptable because of their minimal differential moisture movement (see section 1.7.3).

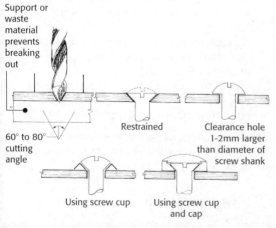

Support or waste material prevents breaking out

Restrained

Clearance hole 1-2mm larger than diameter of screw shank

60° to 80° cutting angle

Using screw cup

Using screw cup and cap

Fig 4.19 Making provision for fixing screws

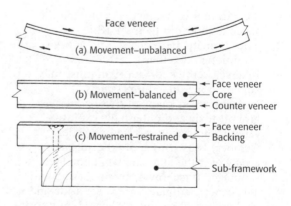

Face veneer

(a) Movement–unbalanced

(b) Movement–balanced   Face veneer / Core / Counter veneer

(c) Movement–restrained   Face veneer / Backing

Sub-framework

Fig 4.20 Stabilising the possible effect of applying a face veneer

Pre-conditioning – a plastics laminate and its support material will, before they can be stuck together, need to be brought into a reasonably balanced condition. Techniques used to bring this about vary, but under normal conditions it can take between three and seven days.

The benefit of pre-conditioning is that it reduces the risk of excessive differential movement between the laminate and its support material, when these are fixed in their final position: bear in mind how the different types of support material can react to the effect of relative humidity changes.

The adhesive used in the bonding process should be similarly conditioned to ensure a uniform temperature at the time of application.

Adhesives – can be considered within four categories:

1   rigid adhesives (thermosetting)
2   urea formaldehyde
    • melamine/urea formaldehyde
    • semi-rigid adhesives -PVAc
3   flexible adhesive (contact adhesives)
4   hot melt adhesives (used only for bonding edge materials).

All should be applied and processed according to their manufacturers' instructions. The most useful of these for on-site work and the small workshop without plate pressing equipment would be contact adhesives. These are either neoprene or natural rubber based and available as solvent or water-based products.

Figure 4.21 shows a procedure for adhering a one face laminate to a stable base board, as follows.

1   The decorative laminate is cut from the sheet as shown in figure 4.18.
2   The laminate, base board, and adhesive are conditioned together.
3   In a well ventilated area (you must fully adhere to manufacturers' safety instructions, as many of these adhesives give off toxic and potentially explosive vapour), an evenly serrated spreader (scraper) is used to apply the adhesive evenly over the laminate while the latter is supported (face down) by the base board.
4   The coated laminate is now set aside and the baseboard is similarly coated, but notice how the spreader (scraper) lines are worked at 90° to those left on the laminate: in this

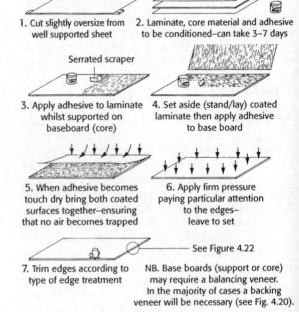

Fine tooth saw or scoring tool

1. Cut slightly oversize from well supported sheet

2. Laminate, core material and adhesive to be conditioned–can take 3–7 days

Serrated scraper

3. Apply adhesive to laminate whilst supported on baseboard (core)

4. Set aside (stand/lay) coated laminate then apply adhesive to base board

5. When adhesive becomes touch dry bring both coated surfaces together–ensuring that no air becomes trapped

6. Apply firm pressure paying particular attention to the edges– leave to set

7. Trim edges according to type of edge treatment

See Figure 4.22

NB. Base boards (support or core) may require a balancing veneer. In the majority of cases a backing veneer will be necessary (see Fig. 4.20).

Fig 4.21   Using a contact adhesive in-situ

way a better bond between the two will be achieved.

5   After the prescribed time (when the adhesive becomes touch dry) the two surfaces are gradually joined together, all the time ensuring that no air becomes trapped between the surfaces, as once they are in final contact there is no recovery.
6   Moderate, yet firm pressure is applied over the whole surface, paying particular attention to the edges.
7   Once the adhesive is set, edge trimming can commence according to the required edge treatment (fig. 4.22).

NB: If a backing veneer is necessary then this is usually applied first before the face veneer, using the same procedure as above.

4.7.3 Edge treatment

Figure 4.22 gives several examples of how door and counter/worktop edges may be finished.

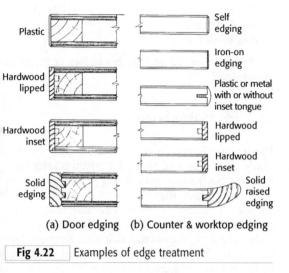

(a) Door edging   (b) Counter & worktop edging

Fig 4.22   Examples of edge treatment

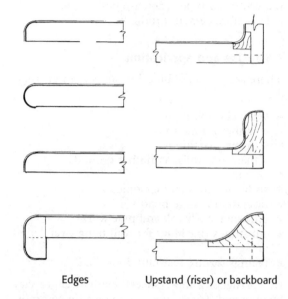

Edges   Upstand (riser) or backboard

Fig 4.23   Examples of post-formed edges and upstands

In the majority of cases machines are used to carry out this task, tools such as portable electric hand routers and edge trimmers (see section 6.15 and fig. 6.42(k & l)), but hand tools must not be ruled out. Hand tools are particularly useful for finishing on site operations and for fine finishing. Very accurate finishes can be achieved with a block plane, flat and shaped files, and cabinet scrapers.

4.7.4 Post-forming

Post-forming is a term used to describe the process involved when bending and bonding specially developed laminates (see table 4.15) over and into various shapes found on the edges and/or surfaces of worktops and sometimes panels – figure 4.23 shows a few different examples.

4.7.5 Check list for when ordering laminated plastics:

● Type sheet – grade to suit application (horizontal [working / decorative] or vertical surface) – table 4.15
● Pattern and colour check
● Sheet *size
● Nominal thickness
● Will a balancing veneer be required
● Availability

*A range of available sheet sizes is given in appendix 2.

4.8   Fibre cement building boards

These boards have always been associated with excellent fire resistance, and, until a few years ago, the main component of that resistance was the now obsolete asbestos binding agent. Because of the serious health risks linked to asbestos fibre, a calcium silicate binder has now replaced it.

The constituents of current boards are largely the result of an interaction of cellulose fibre, lime, cement, silica and a fine protective filler mix during a specialised production process. Probably the greatest advantage that these boards have over other panel products, (with the exception of wood cement particleboard, plasterboard and glass-fibre reinforced gypsum boards) is their excellent fire resistance. However, because of their brittle nature, sheets of fibre cement building boards do require careful handling (see section 4.13).

Form and Appearance – they can take the form of:

● Flat sheets, either smooth, textured, or faced with a rough aggregate.
● Flat weather boarding or planking.
● Corrugated sheets (with various profiles).

Note: Roof slates, shingles, eaves guttering, and

rainwater fall-pipes (down-pipes) are also available as fibre cement products.

4.8.1 Use and application

There are grades of board to suit situations such as:

- vertical cladding to:
  - timber frame walls
  - timber partitions
  - masonry walls via timber grounds
- ceilings
- soffits to roofs and canopies
- fascias and barge boards
- encasing steelwork and pipe work
- substrates (backing) for wall tiling (wet areas)
- facings for fire resistant doors.

By corrugating these sheets, extra strength can be obtained. These sheets have been in common use for many years, covering the roofs of domestic garages, outbuildings and large commercial and industrial structures.

4.8.2 Working and fixing fibre cement products

Sheets can be roughly cut to size either by using a special scoring tool to 'V' the surface then snapping over straight edge, or, with the thicker boards, by using a fine toothed hand saw. Extra hard boards will require specially hardened blades, such as those used in sheet saws to cut sheet metal.

Power tools will need tungsten carbide tipped blades, and provision for dust extraction (see health and safety section 4.14).

Fixing by nails or screws should be undertaken only after first consulting board manufacturers' literature about the correct type, size and fixing pattern (distances from sheet/board edges and between fixings), since these will vary between situations, sheet/board type and thickness.

4.8.3 Check list for ordering Fibre Cement Products:

- Type of sheet – grade to suit application (horizontal or vertical surface).
- Sheet profile if it is to match existing.
- Sheet *size
- Nominal thickness

- Facilities for handling due to its relatively brittle nature
- Availability

*A range of available flat sheet sizes is given in appendix 2.

4.9   Plasterboards

Plasterboards are, in the main, building boards made up of a core of aerated gypsum plaster covered on both faces with heavy sheets of paper. The long edges may be formed and wrapped square or taped to meet the different finish requirements. Such as, a base for plastering a wall, partition, or ceilings, with a traditional plaster finish. Or, as a dry lining material requiring little surface preparation before being able to receive direct decoration.

One of the main advantages of using plasterboard as a lining material, is the fire protection offered by its gypsum core.

Plasterboard Types – there are many types to suit different requirements, for example:

- gypsum wallboard — dry lining material primarily for direct decoration
- gypsum lath — a narrow board to provide a base for plastering
- gypsum plank — a base for plastering, sometimes used to encase steel beams and columns
- gypsum (thistle) baseboard — base for plastering
- gypsum moisture resistant board — plasterboard with added silicon and water repellent paper facing
- gypsum thermal boards — a composite board (see section 4.10) made up of gypsum and expanded polystyrene to increase thermal properties.
- Glasroc (multiboard) – a paperless glass fibre reinforced building board with *similar application to fibre cement sheets*, except that it should not be used externally other than semi-exposed situations.

4.9.1 Application

The plastering trade is usually responsible for fixing plasterboard, but there are occasions where only a small amount is required, the carpenter and joiner may have to carry out this task.

Support material for plasterboard can range from direct fixing with an adhesive to the brick or blockwork, to fixing via timber grounds or timber studs. In the latter case, fixing centres will in the main depend on the thickness of plasterboard and/or specification set down in the working drawings.

All supporting timber should be dried to the required moisture content, and accurately spaced and aligned. The width of the timber supports must be wide enough to provide adequate cover particularly to vertical joints where sheets abut one another.

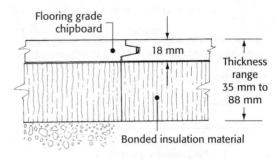

Fig 4.24   Composite flooring material

4.9.2 Fixings

Boards must be fixed according to the manufacturers, instructions. Special plasterboard nails and screws are available from the manufacturers of the board – the correct length and gauge of fixing in relation to board thickness is very important particularly when fixing ceiling boards.

4.9.3 Check list for when ordering plasterboards:

- Type of sheet – type to suit application (horizontal or vertical surface) and fixing location
- Edge profile to suit final finish
- Sheet *size to suit fixing centres and handling
- Nominal thickness
- Paper facing to suit either plaster skim or direct decoration
- Availability of sheets and special fixings, and/or adhesives

*A range of available sheet sizes is given in appendix 2.

4.10 Composite boards

Composite boards are manufactured with special purposes in mind – they can consist of one or more dissimilar materials laminated together as a board. For example, by bonding a mineral wool to the underside of flooring grade chipboard, as shown in figure 4.24, a resilient decking is produced with a built-in means of providing thermal and acoustic properties.

Similarly gypsum wallboards are available with a pre-bonded backing of mineral wool or polystyrene to increase thermal and sound insulation properties whilst also providing added fire protection.

There are many combinations of this nature available to the industry that can enhance the properties of building boards. Some also have the added advantage of pre-bonded decorative surfaces.

4.11 Conditioning wood-based boards and other sheets materials

Nearly all manufactured boards, particular wood based boards and sheets, will require conditioning to the atmosphere to which they will eventually be fixed. This is because of their hygroscopic nature and thereby their inherent ability to take up and shed moisture at will, according to the environment that surrounds them.

In most cases conditioning will simply mean open stacking the materials with air exposure to both faces and as many edges as possible for a pre-scribed period of time, usually a minimum of 48 hours. The temperature and relative humidity must be similar to those where the material will eventually be placed. In this way moisture movement will be allowed to stabilise, thereby avoiding excessive expansion or contraction when the items are fixed in position.

Standard hardboard and tempered hardboard and type L. M. medium board are the only types of manufactured board that may be conditioned with water to accelerate moisture intake. This method of conditioning with water will usually involve rubbing clean water into the back of the

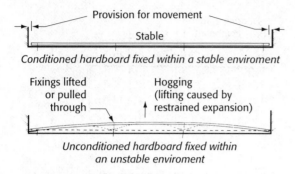

Provision for movement

Stable

Conditioned hardboard fixed within a stable enviroment

Fixings lifted
or pulled
through

Hogging
(lifting caused by
restrained expansion)

*Unconditioned hardboard fixed within
an unstable enviroment*

Fig 4.25   Comparison between using 'conditioned' and 'unconditioned' hardboard

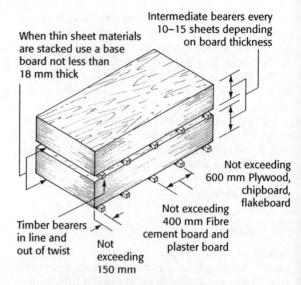

When thin sheet materials
are stacked use a base
board not less than
18 mm thick

Intermediate bearers every
10–15 sheets depending
on board thickness

Not exceeding
600 mm Plywood,
chipboard,
flakeboard

Not exceeding
400 mm Fibre
cement board and
plaster board

Timber bearers
in line and
out of twist

Not
exceeding
150 mm

Note: certain boards with decorative or
film faces may require interleafed face
protection and a means of slide resistance

Fig 4.26   General horizontal storage (except for some decorative boards and sheets)

sheet or board (mesh face) with a brush or mop. Rubbing should start from the centre and the whole surface should be covered. Then, after open stacking wet face to wet face, the material should be allowed to stand for at least 48 hours (tempered hardboard for 72 hours). Failure to condition hardboard, can as shown in figure 4.25, result in it being irreversibly distorted a short time after it has been fixed.

It is important to note however, that sheets and boards treated with flame retardants must never be conditioned with water, or for that matter fixed where they may become damp.

4.12   Storage and stacking

Methods of storage and stacking board and sheet materials, will in the main, depend on their type and location.

Storage – workshop storage may be under controlled conditions within the confines of the workplace. On the other hand on-site provisions may have to be found. In either case what must be borne in mind at all times is as previously stated wherever they are to be housed the atmospheric environment which surrounds them should be as near as possible conducive to that where they will eventually be used.

Stacking – as shown in figures 4.26 and 4.27, providing adequate inline support is provided to prevent the sheets or boards distorting, in many cases both flat and inclined stacking can be employed. Provided the stack is not too high, and boards are accessible from the ground, flat storage is the safest. If inclined storage is employed boards should stand on their long

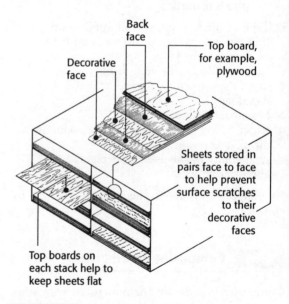

Back
face

Top board,
for example,
plywood

Decorative
face

Sheets stored in
pairs face to face
to help prevent
surface scratches
to their
decorative
faces

Top boards on
each stack help to
keep sheets flat

Fig 4.27   Horizontal compartmentalised storage for laminated plastics and similar thin sheet material

edge and be permanently restrained from toppling over. *Sorting through several reared up board must be strictly prohibited* – each board must be removed one at a time until the selection is made. Those rejected should then be returned one at a time in the same way.

Each type and thickness of board should be stacked separately. This is not only the safest way of storage, but it cuts production cost down dramatically.

4.13   Handling

It goes without saying that all sheet materials must be handled with care, not only to prevent damage to the sheet but more importantly to avoid any risk of personal injury to the handler. The correct lifting procedure must be used at all times and protective clothing worn. Of particular importance here is the use of industrial gloves to prevent splinters from wood-based boards entering the hand, and to prevent cuts from sharp edges, especially from laminated plastics. Feet must also be protected with toe protective footwear. This becomes most important when carrying board material edgeways:

any board dropped edgeways will inflict damage wherever it lands.

Full rigid and semi-rigid sheets/boards, no matter what their thickness, must always be carried and maneuvered by two people (fig. 4.28). If the weight of the board, or perhaps its slipperiness, makes edge carrying difficult, or possibly even dangerous, then it may be worth considering using a 'Porterhook' as a carrying aid. Figure 4.29 shows how a simple hook arrangement can be improvised to make this task both easier and safer. Two single hooks, one at each end, as shown in figure 2.29(a), can be used to carry long sheets with relative ease. A single-handed double hook could be used to handle smaller yet awkward sheets, as shown in figure 4.29(b).

Flimsy sheet material such as laminated plastics must be handled differently. If the sheets are bowed slightly, they become more rigid, this will enable them to be moved more easily. Two people will be required to manipulate full sheets safely.

4.14   Health & safety

Cutting, shaping, forming and drilling any of these products will result in the generation of

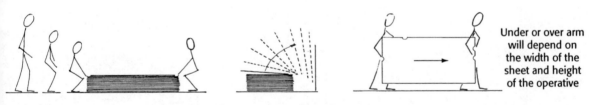

| **Fig 4.28** | Lifting and carrying rigid and semi-rigid sheet material |

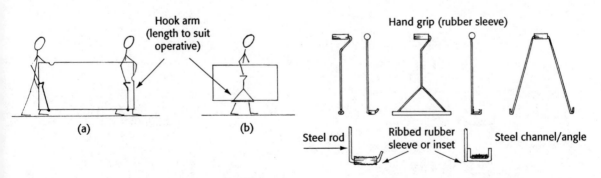

| **Fig 4.29** | The use of purpose made 'porterhooks' |

dust particles that can be harmful to health. Under the Control of Substances Hazardous to Health (COSHH) Regulations, exposure to dust should be prevented or controlled. Limits are set under the Occupational Exposure Limits laid down within Health & Safety at Work Act documentation.

Dust is usually controlled by a means of extraction from individual machines.

Operators employed on site work will be required to use dust masks and goggles for all operations involving cutting, shaping, forming and drilling these materials.

Skin irritation can be an added problem caused by wood, cement or plaster products. Therefore, whenever you are handling newly cut boards, gloves should be considered where appropriate.

In all events the requirements of the Health and Safety at Work Act should be met.

Handtools and Workshop Procedures

5

The starting point for any crafts person must be to master the necessary skills to be able to handle and use hand tools safely and effectively. These skills will, together with the good procedures, usually be reflected in the end product.

5.1 Measuring tools

Measuring tools are used either to transfer measurements from one item to another or for checking pre-stated sizes.

5.1.1 Scale rule

At some stage in your career you will have to take sizes from or enter sizes on to a drawing – therefore you must familiarise yourself with methods of enlarging or reducing measurements accordingly. It is essential to remember that all sizes stated and labelled on working drawings, will be true full sizes, but for practical reasons these sizes will, in some cases, have to be proportionally reduced to suit various paper sizes by using one of the following scales: 1:2 (half full size), 1:5, 1:10, 1:20, 1:50, 1:100, 1:200, 1:1250, 1:2500.

Figure 5.1 illustrates the use of a scale rule, which enables lengths measured on a drawing to be converted to full-size measurements and vice versa.

5.1.2 Four-fold metre rule

This rule should have top priority on your list of tools. Not only is it capable of accurate measurement, it is also very adaptable (see fig. 5.2). It is available in both plastics and wood, and cal-

ibrated in both imperial and metric units. Some models (clinometer rules) also incorporate in their design a spirit-level and a circle of degrees from 0° to 180°.

With care, these rules will last for many years. It is therefore important when choosing one to find the type and make that suits your hand. Ideally, the rule should be kept about your person while at work. The most suitable place while working at the bench, or on site at ground level is usually in a rule slide pocket sewn to the trouser leg of a bib and brace overall etc. The use of a seat or back pocket is not a good idea.

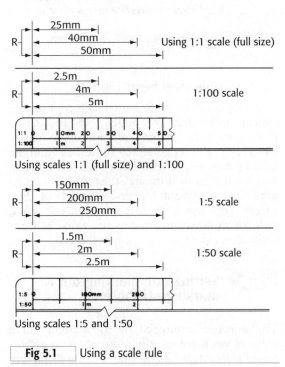

| **Fig 5.1** | Using a scale rule |

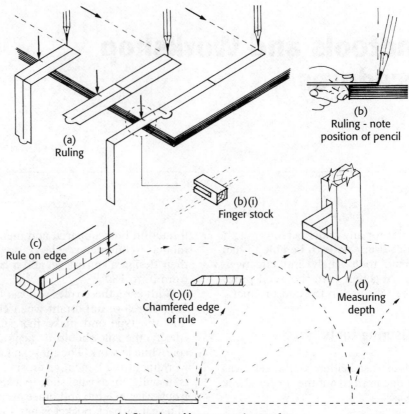

(a)
Ruling

(b)
Ruling - note
position of pencil

(b)(i)
Finger stock

(c)
Rule on edge

(c)(i)
Chamfered edge
of rule

(d)
Measuring
depth

(e) Stepping - Measurements over 1m
(not an accurate means of measurement, only used to give an approximate reading)

Fig 5.2   Versatility of a four fold metre rule

5.1.3 Flexible steel tapes (fig 5.3)

These tapes retract on to a small enclosed spring-loaded drum and are pulled out and either pushed back in or have an automatic return which can be stopped at any distance within the limit of the tape's length. Their overall length can vary from 3 m to 8 m, and they usually remain semi-rigid for about the first 500 mm of their length. This type of tool is an invaluable asset, particularly when involved in site work, as it fits easily in the pocket or clips over the belt.

5.2   Setting-out, marking-out & marking-off tools

The drawings produced by the designer of a piece of work are usually reduced to an appropriate scale, (fig. 5.1) so that an overall picture

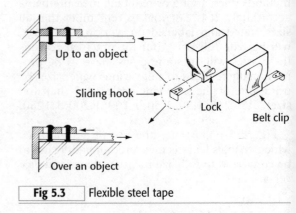

Up to an object

Sliding hook

Lock

Belt clip

Over an object

Fig 5.3   Flexible steel tape

may be presented to the client. Once approval has been given, the setting-out programme can begin. This will involve redrawing various full-size sections through all the components necessary for the construction, to enable the joiner to visualise all the joint details etc. and make any adjustments to section sizes.

5.2.1 Setting-out bench equipment

Setting-out is done on what is known as a rod. A rod may be a sheet of paper, hardboard, plywood, or a board of timber. By adopting a standard setting-out procedure, it is possible to simplify this process. For example (see fig. 5.4):

- draw all sections with their face side towards you (fig. 5.4(a)),
- draw vertical sections (VS) first – with their tops to your left (fig. 2.5(b)),
- draw horizontal sections (HS) above VS – keeping members with identical sections (top rail & stile) in line on the HS and VS, for example, top rail 2.5(b) with the stile figure 2.5(c),
- allow a minimum of 20 mm between sections (fig. 2.5(d));
- dimension only overall heights, widths, and depths.

Setting-out will involve the use of some, if not all, of the following tools and equipment;

- a scale rule (fig. 5.1),
- a straight-edge (400–1000 mm),
- a four-fold metre rule, figure 5.2 shows its versatility – see also figure 5.4,
- precision metal rules – available from 150–1200 mm in length,
- drafting tape, drawing-board clips, or drawing pins,
- an HB pencil,
- a try-square – figure 5.9 shows its application – also see figure 5.4,
- a combination square – figure 5.9 shows its application – also see figure 5.4,
- dividers – figure 5.5 shows various methods of use,
- compasses (fig. 5.5) – can be used as dividers,
- a trammel (fig. 5.6) – used with a trammel bar for scribing large circles or arcs etc. – one of the scribers can be replaced with a pencil.

Figure 5.4 shows a space left on the rod for a cutting list. This is an itemised list of all the material sizes required to complete a piece of work. A typical simplified format for a cutting list is shown in figure 5.7, together with provision for items of hardware (ironmongery) etc. It follows that the information with regard to timber sizes and quantity, will be required by the wood machinist (Chapter 9).

Marking-off involves the transfer of rod dimensions on to the pieces of timber and/or other materials needed. Provided that the rod is correct (*always double check*), its use (see fig. 5.8)

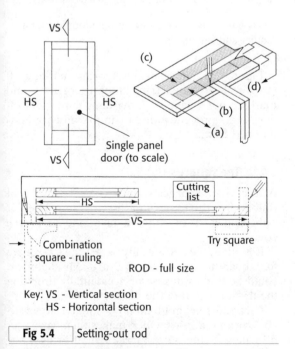

Key: VS - Vertical section
     HS - Horizontal section

Fig 5.4   Setting-out rod

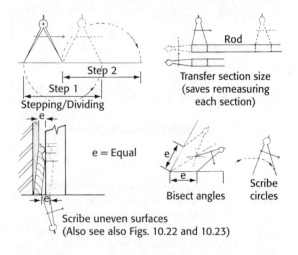

Fig 5.5   Using dividers or compass

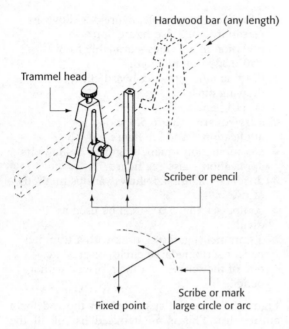

Fig 5.6 | Trammel heads and bar

Job title						
Quantity		Saw size (Ex)		Finish size		
No.	Item	L	W	T	W	T	Remarks

(second header/body table)

Job title						
Quantity		Saw size (Ex)		Finish size		
No.	Item	L	W	T	W	T	Remarks

Sundries - Nails, screws, hardware, etc.	
Quantity	Description

L = length, W = width, T = thickness

Fig 5.7 | Cutting and hardware (ironmongery) list

reduces the risk of duplicating errors, especially when more than one item is required.

Once all the material has been reduced to size (as per the cutting list) and checked to see that its face side is not twisted and that all the face edges are square with their respective face sides, the marking-off process can begin.

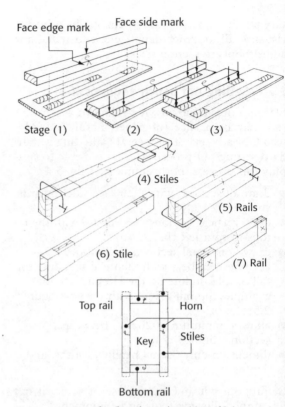

Stages 1-2-3 Transfer (lengths and position of joints, etc.)
   4-5 Coupled transfer to ensure identical pairs
       (material must be held firmly by vice or cramp)
Stages   6-7 Divide - mark each face to indicate joints, etc.

Fig 5.8 | Typical marking-off/out procedure for a mortice and tenoned frame

Figure 5.8 illustrates a typical marking-off (stages 1–3 transferring measurements) and marking-out (stage 4–7 marking joints, cut-off lines etc.) procedure for a simple mortise-and-tenoned frame.

5.2.2 **Try-squares** (fig 5.9)

As their name suggests, these (try) test pieces of timber for squareness or for marking lines at right angles from either a face side or a face edge.

It is advisable periodically to test the try-square for squareness (see fig. 5.9(a)). Misalignment could be due to misuse or accidentally dropping the try-square on to the floor.

Large all-steel graduated try-squares are available without a stock which makes them very useful when setting out on large flat surfaces. They

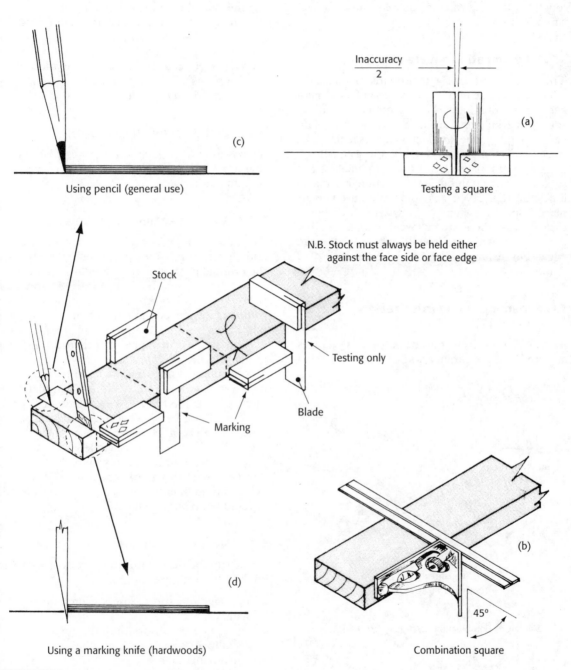

(c)

Using pencil (general use)

Inaccuracy
___2___

(a)

Testing a square

N.B. Stock must always be held either
against the face side or face edge

Stock

Testing only

Blade

Marking

(d)

Using a marking knife (hardwoods)

45°

Combination square

(b)

Fig 5.9   Using a try square

look like the steel roofing square – these are featured in book 2.

5.2.3 Combination square (fig. 5.9(b))

This can be used as a try-square, but has the added advantage of being very versatile, in that it has many other uses. For example, as a mitre square, marking and testing angles of 45°, height gauge, depth gauge, marking gauge (see fig. 5.4), spirit-level (some models only), and as a rule.

It is common practice to use a pencil with a square, as shown in fig 5.9(c). Although a marking knife (fig. 5.9(d)) is sometimes used in its place (especially when working with hardwoods) to cut across the first few layers of fibres so that the saw cut which follows leaves a sharp clean edge. For example, at the shoulder line of a tenon.

5.2.4 Marking and mortise gauges (fig. 5.10)

As can be seen from the diagram, these gauges are similar in appearance and function, i.e. scor-

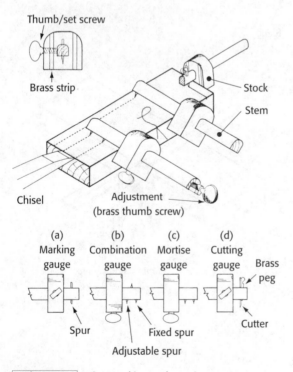

Thumb/set screw

Brass strip

Stock

Stem

Chisel

Adjustment (brass thumb screw)

| (a) | (b) | (c) | (d) |
| Marking gauge | Combination gauge | Mortise gauge | Cutting gauge |

Brass peg

Spur

Fixed spur

Adjustable spur

Cutter

Fig 5.10   Using marking and mortice gauge

ing lines parallel to the edge of a piece of timber. The main difference is that the marking gauge (fig 5.10(a)) scores only a single line but the mortise gauge (fig 5.10(c)) scores two in one pass. It is possible to buy a combination gauge that can perform both operations simply by being turned over (see fig. 5.10(b)).

5.2.5 Cutting gauge (fig. 5.10(d))

This is used to cut across the fibres of timber. It therefore has a similar function to that of a marking knife.

5.2.6 Marking angles and bevels

The combination square (previously mentioned, and shown in fig. 5.9) has probably superseded the original mitre square, which looked like a set square but had its blade fixed at 45° and 135° instead of 90°. However, two of the most useful pieces of bench equipment when dealing with mitres, are a mitre template and a square and mitre template. Examples of their use are shown in figure 5.11.

Any angles other than 45° will have to be transferred with the aid of either a template premarked from the rod or the site situation, or by using a sliding bevel. This has a blade that can slide within the stock and be locked to any angle (fig. 5.12). The bevel as a whole can sometimes prove to be a little cumbersome for marking dovetail joints on narrow boards. This can be overcome quite easily by using a purpose-made dovetail template (see fig. 5.12).

5.2.7 Site measuring tools

These can be dealt with within related topics in book 2. For example:

- long tape measures – measuring long distances
- site squares – setting-out large angles (usually 90°) using as a template, or optical instrument.
- levels:
  - spirit levels – for checking lines both horizontally and vertically.
  - water levels – setting-out and checking lines and heights horizontally.
  - optical levels – setting out and checking lines and heights horizontally.

Square over moulded section

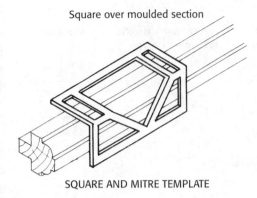

SQUARE AND MITRE TEMPLATE

Score mitre profile with chisel or
marking knife held flat

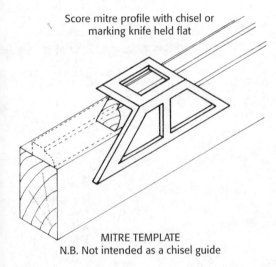

MITRE TEMPLATE
N.B. Not intended as a chisel guide

Fig 5.11 Marking around moulded sections prior to
cutting

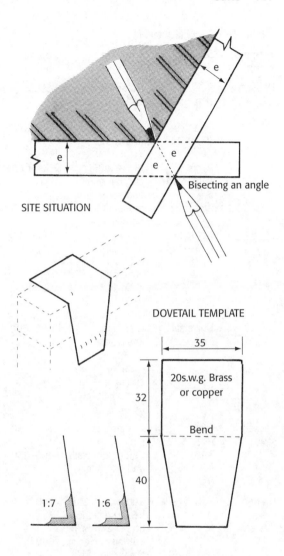

SITE SITUATION

Bisecting an angle

DOVETAIL TEMPLATE

35

20s.w.g. Brass
or copper

32

Bend

40

1:7   1:6

Blade

SLIDING BEVEL

e

Locking device

Stock

e

e = Equal width or angle

Fig 5.12 Aids for marking and transferring anglers

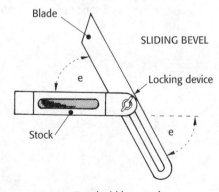

- laser levels – setting out and checking
lines and heights horizontally.

5.3   Saws

Saws are designed to cut both along and across
the grain of wood (except the rip saw – see table
5.1), and the saw's efficiency will be determined
by:

- the type and choice of saw,
- the saw's condition,
- the application,
- the material being cut.

Table 5.1   Saw fact sheet

Type or group	Saw	Function	Blade length (mm)	Teeth shape	Teeth per 25 mm	Handle type	Remarks
Handsaws	Rip (Fig. 5.13)	Cutting wood with the grain (ripping)	650	3° 60°	4–6		Seldom used below six teeth per 25 mm
	Cross cut (Fig. 5.14)	Cutting wood across the grain	600 to 650		7–8		Can also he used fir rip sawing
	Panel (Figs. 5.15/ 5.16)	Cross-cutting thinner wood and some manufactured board	500 to 550	14° 60°	10		Easy to use handle
	Hardpoint (crosscut)	Can be used for cutting with and across the grain	500–600		7–10		Hardpoint - non-sharpenable with extended cutting life – up to 6–7 longer than sharpenable blades.
	Universal saw	Cuts along and across the grain – ideal for cutting manufactured board	500–600		7–10		
Backed saws	Tenon (Figs. 5.17/ 5.18), and dovetail (Fig. 5.19)	Tenons and general work	300–450		12–14		Depth of cut restricted by back strip (blade stiffener)
		Cutting dovetails and fine work	200–250		18–20		
Framed saws	Bow	Cutting curves in heavy sectioned timber and M/B	200 to 300		12 ±		Radius of cut restricted by blade width
	Coping (Fig. 5.20)	Cutting curves in timber and M/B	160		14		Thin narrow blade
	Hacksaw (Fig. 5.21)	Cutting hard and soft metals	250 to 300		14–32		Small teeth - to cut thin materials. The larger the teeth, the less liable to clog – small frame hacksaw (Fig. 5.22)
Narrow-blade saws	Compass	Cutting slow curves in heavy, large work	300 to 450		10 ±		Interchangeable blades of various widths – unrestricted by a frame
	Pad or key hole	Enclosed cuts - piercing panels, etc	200 to 300		10 ±		Narrow blade partly housed in a handle, therefore length adjusts

Note: M/B – manufactured boards

5.3.1 Selection

Broadly speaking, saws can be categorised into four groups;

i   handsaws,
ii   backed saws,
iii framed saws,
iv   narrow-blade saws.

As can be seen from table 5.1, each type/group can be further broken down into two or three specifically named saws, which are available in a variety of sizes and shapes to suit particular functions.

5.3.2 Condition of saw

It is important that saws are kept clean (free from rust) and sharp at all times (see section 5.13.2). Dull or blunt teeth not only reduce the efficiency of the saw, but also render it potentially dangerous. For example, insufficient set could cause the saw to jam in its own kerf and then buckle, or even break (see fig. 5.23).

5.3.3 Methods of use

The way the saw is used will depend on the following factors:
a   the type and condition of the wood being cut,
b   the direction of cut – ripping or cross-cutting,
c   the location – bench work or site work.

Practical illustrated examples are shown in figures 5.13 to 5.22.

Note: *the emphasis on safety, i.e. the position of hands and blades and body balance etc.*

5.3.4 Materials being cut

A vast variety of wood species are used in the building industry today, and many, if not all, will at some time be sawn by hand. The modern saw is ideally suited to meet most of the demands made upon it, although there are instances

Fig 5.13 Ripping with a hand saw (notice the position of the fore finger and how the arm and shoulder is in line with the saw blade)

Fig 5.14 Cross cutting with a hand saw (notice the position of the fore finger and how the arm and shoulder is in line with the saw blade)

Fig 5.15   Sawing down the grain (vertically) with a panel saw

Fig 5.16   Sawing down the grain with the material angled – two saw lines are visible

Fig 5.17a   Tenon saw – starting a cut, vice held

Fig 5.17b   Tenon saw – starting a cut using a bench hook

Fig 5.18   Sawing down the grain with a tenon saw

Fig 5.19   Dovetail saw – starting a cut

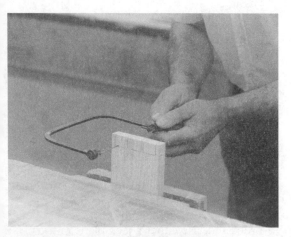

Fig 5.21   Using a standard hacksaw

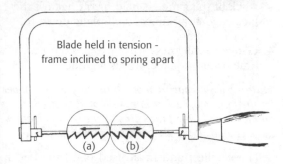

Blade held in tension -
frame inclined to spring apart

(a)   (b)

(a) To cut on a forward stoke will tend to bend or break th blade
(b) To cut on a back stroke will tend to keep blade taut

Fig 5.20   Using a coping saw

Fig 5.22   Using a junior hacksaw

Saw set provides clearance–
prevents saw from binding in kerf

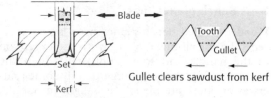

Blade

Tooth

Gullet

Set

Kerf

Gullet clears sawdust from kerf

Sawcut referred to as kerf

Fig 5.23   Providing saw blade clearance

where it will be necessary to modify general saw-ing techniques, for example when dealing with wood that is:

a   *Very hard* – use a medium cut hardpoint saw.
b   *Of very high moisture content* – use a tapered-back saw with slight increase in set.

c   *Extremely resinous* – use a tapered-back saw – keep blade clean of resin residue.
d   *Case-hardened* – keep the kerf (see fig. 5.23) open, possibly be using small wedges.

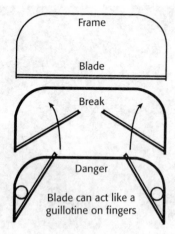

When using frames saws–especially hacksaws–
always keep fingers outside the framework in
case the blade breaks–see Fig 5.20, 21, 22

| **Fig 5.24** | Frame saw safety |

5.4   Planes

There are many types of plane. All are capable of
cutting wood by producing shavings, but not all
are designed to produce *plane* flat surfaces as the
name implies. However, as can be seen from
table 5.2, each plane has its own function and,
for the sake of convenience, planes have been
placed within one of two groups;

1   bench planes,
2   special purpose planes.

5.4.1 Bench planes (figs 2.25–2.28)

Wooden-bodied bench planes have been superseded by the all-metal plane (with the exception
of the handle and front knob). Although the
wooden jack-plane is still regarded by some joiners as an ideal site plane – it is light to handle,
and less liable to break if dropped. Probably the
greatest asset of the wooden jack-plane is its
ability to remove waste wood rapidly – where
accuracy is not too important.

All metal bench planes are similarly constructed with regard to blade angle (45°), adjustment, and alignment (fig. 5.39); variations are
primarily due to the size or design of the plane

sole (fig 5.44), which determines the function
(table 5.2).

Figures 5.40 and 5.41 show a jack-plane being
used for flatting and edging a short piece of timber. *Notice particularly the position of the hands in
relation to the operation being carried out.*

Processing a piece of sawn timber by hand is
carried out as follows:

1   Select, and using a jack-plane, plane a face
side (best side) straight (fig. 5.42(a)) and out
of twist (fig. 5.42(b)). *(Winding laths are used
to accentuate the degree of twist.)* Label the
side with a face-side mark (fig. 5.9).
2   Depending on the length of the timber being
processed, use either a jack plane or a try-
plane (fig. 5.43) to plane a face edge straight
and square to the face side (keep checking
with a try-square). As can be seen from
figure 5.44 the shorter the sole the more
difficult it is to produce a long straight edge.
A long length of timber will require end
support to prevent it tipping – this can be
achieved by positioning a peg in one of a
series of pre-bored holes in the face or leg of
the bench, see figure 5.45. On completion,
credit the edge with a face-edge mark (fig.
5.9).
3   Gauge to width (fig. 5.10), ensuring that the
stock of the gauge is held firmly against the
face edge at all times, then plane down to
the gauge line.
4   Gauge to thickness – as in step 3 but this
time using the face side as a guide.

Note: *when preparing more than one piece of timber
for the same job, each operation should be carried out
on all pieces before proceeding to the next operation,
i.e. face side all pieces, face edge all pieces, and so on.*

The smallest and most used of all the bench
planes is the smoothing plane. This is very easy
to handle and, although designed as a fine finishing plane for dressing joints and surfaces
alike, it is used as a general-purpose plane for
both bench and site work. Figure 5.46 shows a
smoothing plane being used to dress (smooth
and flat) a panelled door, and how by tilting the
plane it is possible to test for flatness (this applies
to all bench planes) – the amount and position of
the light showing under its edge will determine
whether the surface is round or hollow.

Note: *the direction of the plane on the turn at corners or rail junctions – for example working from stile*

Table 5.2 Plane fact sheet

Group	Plane	Function	Length (mm)	Blade width (mm)	Remarks
Bench planes	Smoothing (Fig. 5.25a)	Finishing flat surfaces	*240, *245, 260	45, 50, 60	50 mm the most common blade width
	Record CSBB (Fig. 5.25b)	As above	245	60	'Norris' type cutter adjustment – combining depth of cut with blade adjustment
	Jack (Fig. 5.26)	Processing saw timber	*355 380 mm	50, 60	60 mm the most common blade width
	Fore Jointer (try plane) (Fig. 5.27)	Planing long edges (not wider than the plane's sole) straight & true.	*455 *560 610	60	The longer the sole, the greater the degree of accuracy
	Bench rebate (carriage or badger) (Fig. 5.28)	Finishing large rebates	235, 330	54	Its blade is exposed across full width of sole
Special planes	Block (Fig.5.30)	Trimming – end grain	140, 180, 205	42	Cutter seats at 20° or 12° (suitable for trimming laminated plastics, depending on type
	Circular (compass plane) (Fig. 5.31)	Planing convex or concave surfaces	235, 330	54	Spring-steel sole adjusts from flat to either concave or convex curves
	Rebate (Fig. 5.32)	Cutting rebates with or across the grain	215	38	Both the width and depth of rebate are adjustable
	Shoulder/rebate (Fig. 5.33)	Fine cuts across grain, and general fine work	152, 204	18, 25, 29, 32	Some makes adapt to chisel planes
	Bullnose/shoulder rebate (Fig.5.34)	As above, plus working into confined corners	100	25, 29	
	Side rebate (Fig. 5.35)	Widening rebates or grooves – with or across grain	140		Removable nose - works into corners. Double-bladed - right and left hand. Fitted with depth gauge and/or fence.
	Plough (Fig. 5.36)	Cutting grooves of various widths and depths – with and across grain	248	3 to 12	Both width and depth of groove adjustable
	Combination	As above, plus rebates, beading, tongues, etc	254	18 cutters, various shapes and sizes	Not to he confused with a multi plane, which has a rang 24 cutters
	Open-throat router (Fig. 5.37)	Levelling bottoms of grooves, trenches, etc.		6, 12 and V	A fence attachment allows it to follow straight or round edges
	Spokeshave (Fig. 5.38) flat bottom & round bottom types	Shaving convex ot concave surfaces - depending on type see Fig. 5.51	250	54	Available with or without 'micro blade depth adjustment

Note: * available with corrugated soles (Fig. 5.29) which are better when planing resinous timber.

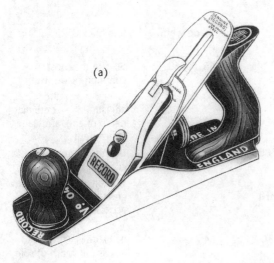

(a)

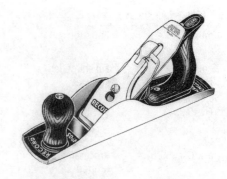

Fig 5.26 Jack plane (kind permission from Record tools limited)

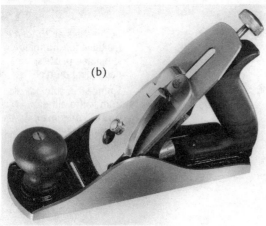

(b)

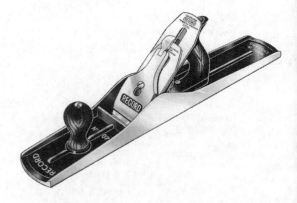

Fig 5.27 Jointer or try-plane (with kind permission from Record tools limited)

Fig 5.25 Smoothing planes (with kind permission from Record tools limited)

to rails or rail to stile – will be determined by the direction of the wood grain.

5.4.2 Special planes (figs 5.30–5.38)

It should be noted that there are more special planes than the ten planes listed in table 5.2, and there are also variations in both style and size.

The kinds of planes selected for your tool kit will depend on the type of work you are employed to do. There are, however, about four types of plane which, if not essential in your work, you should find very useful. They are the:

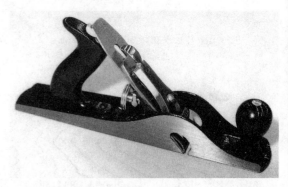

Fig 5.28 Bench rebate plane (carriage or badger plane) (with kind permission from Stanley tools limited)

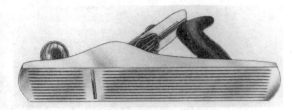

Fig 5.29 Corrugated sole (with kind permission from Record tools limited)

Fig 5.32 Rebate plane (with kind permission from Record tools limited)

Fig 5.30 Block plane (with kind permission from Record tools limited)

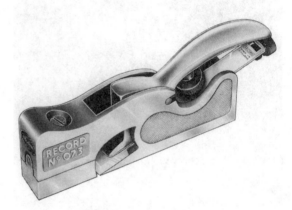

Fig 5.33 Shoulder/rebate plane (with kind permission form Record tools limited)

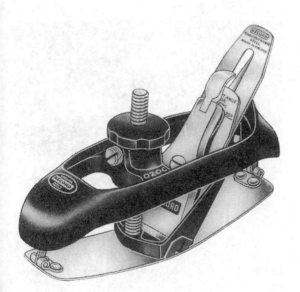

Fig 5.31 Circular compass plane (with kind permission from Record tools limited)

a block plane (fig. 5.30),
b rebate plane (fig. 5.32),
c plough plane (fig. 5.36),
d spokeshaves (fig. 5.38).

Block plane (fig. 5.30) – capable of tackling the most awkward cross-grains of both hardwood and softwood, not to mention the edges of manufactured boards and laminated plastics. Some models have the advantage of an adjustable mouth (fig. 5.119) and/or their blade set to an extra low angle of 12°, such a combination can

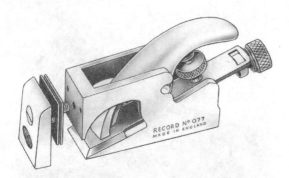

Fig 5.34   Bullnose/shoulder/rebate plane (with kind permission from Record tools limited)

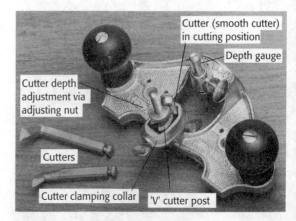

Fig 5.35   Side rebate plane in the use (with kind permission from Stanley tools limited)

Fig 5.36   Plough plane (with kind permission from Record tools limited)

Fig 5.37   Open – throat router and its use (with kind permission from Stanley tools limited)

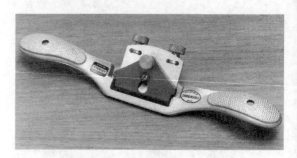

Fig 5.38   Spokeshave (with kind permission from Stanley tools limited)

increase cutting efficiency. Block planes are designed to be used both single and double-handed.

Rebate plane (fig. 5.32) – figure 5.47 shows a rebate plane being used to cut a rebate of controlled size, by using a width-guide fence and a

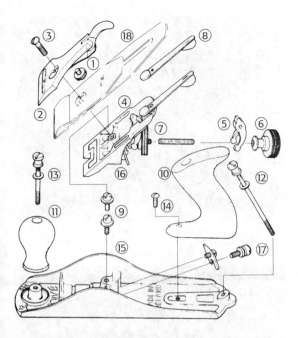

Fig 5.39 Exposed view of a Stanley bench plane (with kind permission from Stanley tools limited) **1** – cap iron screw; **2** – lever cap; **3** – lever screw; **4** – frog, complete; **5** – 'Y' adjusting lever; **6** – adjusting nut; **7** – adjusting nut screw; **8** – LA (lateral adjustment) lever; **9** – frog screw and washer; **10** – plane handle; **11** – plane knob; **12** – handle screw and nut; **13** – knob screw and nut; **14** – handle toe screw; **15** – plane bottom, sole; **16** – frog clip and screw; **17** – frog adjusting screw; **18** – cutting iron (blade) and cap iron

Fig 5.40 Flatting – planing the face of a piece of timber placed against a bench stop

depth stop. It is, however, very important that the cutting face of the plane is held firmly and square to the face side or edge of the timber throughout the whole operation.

Because the blade of a rebate plane has to extend across the whole width of its sole, a cut finger can easily result from careless handling. Particular care should therefore be taken to keep fingers away from the blade during its use, and especially when making the desired depth and/or width adjustments.

Plough plane (fig. 5.36) – figure 5.48 shows a plough plane cutting a groove in the edge of a short length of timber. It is also capable of cutting rebates to the cutter widths provided and by the method shown in figure 5.49.

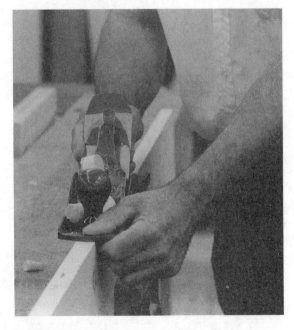

Fig 5.41 Edging – planing the edge of a piece of timber held in a bench vice with finished face positioned towards the operative – four fingers acts only as a guide to centralise the plane during the operation – not as a fence

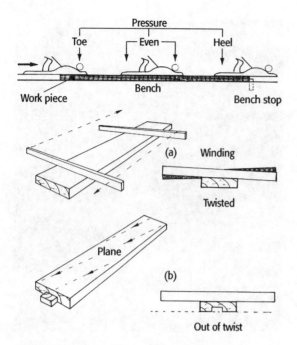

N.B. Winding laths must be parallel in their length

Fig 5.42   Preparing a face side

The method of applying the plane to the wood is common to both rebate and plough planes, in that the cut should be started at the forward end and gradually moved back until the process is completed (see fig. 5.50). Joiners often prefer to use larger flat-bottomed planes for shaping convex curves, but the efficiency and ease of operation of a flat-bottomed spokeshave can only be realised when the technique of using this tool has been mastered. Figure 5.51 shows a spokeshave in use – note particularly the position of the thumbs and forefingers, giving good control over the position of the blade in relation to its direction of cut, always with the grain.

5.5   Boring tools

Boring tools are those tools that enable the cutting of circular holes of a predetermined size into and below the surface of a given material. They can be divided into three groups;

Fig 5.43   Using a try plane (jointer)
a   toe down – to middle
b   even pressure – self weight of plane
c   heal down to end

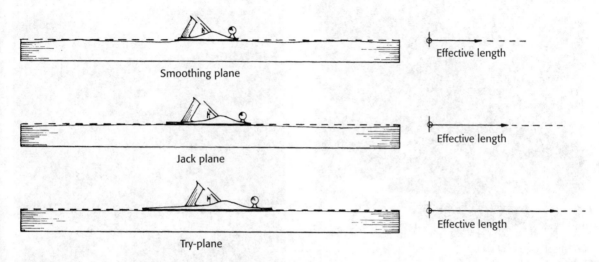

Fig 5.44 Longer the sole – greater the accuracy for a straight edge

Fig 5.45 Providing end support via bench pegs

1 standard bits,
2 special bits,
3 drills.

Table 5.3 has classified the above under the headings of:

a type,
b function,
c motive power,
d hole size,
e shank section.

(Table 5.3 is meant to be used in conjunction with the illustrations of tools in figure 5.52.)

The driving force necessary for such tools to operate is provided by hand, by electricity (chap 6 & 7), or by compressed air.

Hand operations involve the use of a bit or drill-holder (chuck), operated by a system of levers and gears. There are two main types; the *carpenter's brace* and the hand drill, also known as the *wheelbrace*.

5.5.1 Carpenter's brace

This has a two-jaw chuck, of either the *alligator* or the *universal* type. The alligator type has been designed to take square-tapered shanks, whereas the universal type takes round, tapered and straight, as well as square tapered shanks. The amount of force applied to the bit or drill will depend largely on the *sweep* of the brace. Figure 5.53 shows how the style and sweep can vary.

There are three main types of brace:

i **Ratchet brace** – the ratchet mechanism allows the brace to be used where full sweeps are restricted (an example is shown in fig. 5.54). It also provides extra turning power (by eliminating overhand movement), so often needed when boring large-diameter holes or using a turn-screw bit.

ii **Plain brace** (non-ratchet type) – limited to use in unrestricted situations only, *so not to be recommended.*

iii **Joist brace** (upright brace) – for use in awkward spaces, for example between joists.

Method of use – probably the most difficult part of the whole process of boring a hole is keeping the brace either vertical or horizontal to the workpiece throughout the whole operation. Accuracy depends on all-round vision, so, until you have mastered the art of accurately assessing horizontally and verticality, seek assistance.

(a) Smoothing a door stile

(c) Smoothing a door rail

(b) Dressing flat a joint between door stile and rail – a slow turning action towards the joint helps to avoid unnecessary tearing of the grain

(d) Testing a corner joint for flatness by tilting the plane – notice the light showing between the edge of the plane sole and the surface of the timber, indicating that, at this position, the surface is hollow

Fig 5.46   Using a smoothing plane to dress the faces of a fully assembled, glued, cramped and set panelled door

Figures 5.55 and 5.56 show situations where the assistant directs the operative by simple hand signals. Note also, that for vertical boring the operative's head is kept well away from the brace, giving good vision and unrestricted movement to its sweep – only light pressure should be needed if the bit is kept sharp. Horizontal boring support is given to the brace by arm over leg as shown –

Fig 5.47 | Planing a rebate

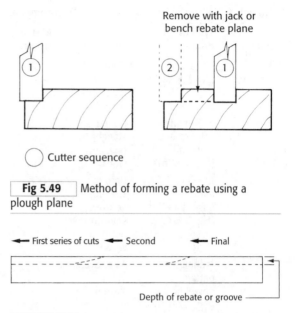

Remove with jack or
bench rebate plane

Cutter sequence

Fig 5.49 | Method of forming a rebate using a plough plane

← First series of cuts ← Second → Final

Depth of rebate or groove

Fig 5.50 | Method of applying a rebate or plough plane to the work piece

in this way good balance can be achieved and maintained throughout the process. The stomach should *not* be used as a form of support.

The use of an upturned try-square etc. placed on the bench, as a vertical guide *is potentially dangerous, any* sudden downward movement, due to the bit breaking through the workpiece, could result in an accident.

Figure 5.57 shows the use and limitations of some of the bits and drills mentioned, together with two methods of breaking through the opposite side of the wood without splitting it, i.e. reversing the direction of the bit or temporarily securing a piece of waste wood to the point of breaking through.

5.5.2 **Hand-drill** (wheel brace) (fig. 5.58)

This has a three-jaw self-centring chuck, designed specifically to take straight-sided drills. It is used in conjunction with twist drills or masonry drills.

5.5.3 **Boring devices and aids**

Probably the most common of all boring devices is the bradawl (fig. 5.59) – a steel blade fixed into a wood or plastics handle – used mainly to bore pilot holes for screw threads.

Fig 5.48 | Cutting a groove with a plough plane

Table 5.3   Characteristics of bits and drills in common use. (Not all types are available in metric sizes.)

Group	Type of bit/ drill	Function	Motive power	Range of common hole sizes	Shank section	Remarks
Bits (standard)	Centre bit (Fig. 5.52(a))	Cutting shallow holes in wood*	CB	¼–2¼ in	❑	
	Irwin-pattern solid-centre auger bit (Fig. 5.22(b))	Boring straight holes in wood*	CB	¼–1½ in 6–38 mm	❑	General purpose bit
	Jennings-pattern auger bit (Fig. 5.22(c))	Boring straight, accurate, smooth holes in wood*	CB	¼–1½ in	❑	
	Jennings-pattern dowel bit (Fig. 5.22(d))	As above, only shorter	CB	⅜ and ½ in-only	❑	Used in conjuction with wood dowel
	Combination auger bits (Fig. 5.22(e))	Cutting very clean holes in wood	CB ED	¼–1¼ in 6–32 mm	○	Must only be used with slow cutting speeds
	Countersink (Fig. 5.22(f))	Enlarging sides of holes	CB ED	⅜, ½, ⅝ in	❑○	Rose, shell (snail) heads available depending on material being cut and speed
Bits (special)	Expansive (expansion) bit (Fig. 5.22(g))	Cutting large shallow holes in wood*	CB	⅞–3 in	❑	Adjustable to any diameter within its range
	Scotch-eyed auger bit (hand) (Fig. 5.22(h))	Boring deep holes in wood	T	¼–1 in		Turned by hand via 'T' bar. Length determined by diameter of auger
	Forstner bit (Fig. 5.22(i))	Cutting shallow bottomed holes in wood*	CB ED	⅜–2 in	❑○	Ideal for starting a stopped housing
	Dowel-sharpener bit	Chamfering end of dowel	CB		❑	Pointed dowel – aids entry into dowel holes

Table 5.3 (Continued)

Group	Type of bit/ drill	Function	Motive power	Range of common hole sizes	Shank section	Remarks
Bits (special) (continued)	Turn-screw bit (Fig. 5.22(j))	Driving large screws	CB	¼, ³⁄₁₆, ³⁄₈, ⁷⁄₁₆ in	❏	Very powerful screwdriver
	Flat bit (Fig. 5.22(k))	Bores holes in all forms of wood quickly, cleanly	ED	¼–1½ in 6–38 mm	○	Deep holes may wander Extension shank available
	Screw bit (Screwmate) (Fig. 5.22(li))	Drills pilot, clearance and countersink in one operator	ED	Screw size 1 in × 6 to 1½ in × 10 25 mm × 6 to 38 mm × 10	○	Saves time changing different bits
	Screw sink (Fig. 5.22(lii))	Combination counterbore all-in-one boring of screw hole and plug hole	ED	¾ in 3 6 to 2 in 3 12	○	Depth of counterbore can be varied. Use with Stanley plug cutter
	Plug cutter (fig. 5.22(m))	Cuts plugs to fill counterbored holes	ED	To suit screw gauges 6 to 12 and hole sizes ³⁄₈, ½, ⁵⁄₈ in	○	Use only with a drill fixed into a bench drill stand, and with the wood cramped down firmly (see manufacturer's instructions for maximum safe speed.)
Drills	Twist drills	Boring wood*, metals, plastics	HD ED	¹⁄₁₆ to ½ in 1–13 mm	○	A few sizes available as ❏
	Masonry drills (tungsten-carbide)	Boring masonry, brickwork, concrete	HD ED	Numbers 6–20 ⁵⁄₁₆ to ³⁄₈ in†	○	Available for both rotary and impact (percussion-action) drills

Note: † Larger diameter drills are available; CB – carpenter's brace; HD – hand drill (wheel-brace); ED – electric drill (N.B. Bits used in ED must never have screw point other than with a very slow cutting speed and adjustable torque mechanism); ❏ – square-tapered shank; ○ – straight rounded shank; * wood and all wood based products; ⬠ – hexagonal end shank

Flat-bottomed sole   Rounded sole

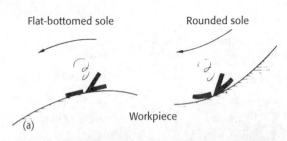

Workpiece

(a)

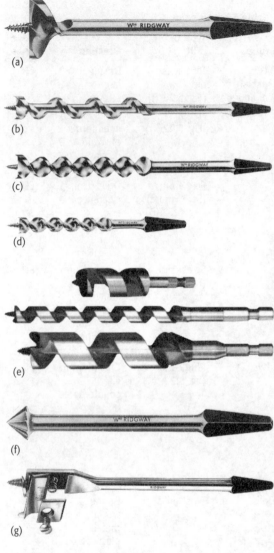

Fig 5.51   Spokeshave in use
a   only make cuts in the direction of the wood grain
b   using a flat bottomed spokeshave to cut over the
    curve
c   using a rounded bottomed spokeshave to make a cut
    into a hollow

Fig 5.52   A range of bits sinks and cutters (not to
scale)
a   centre bit
b   Irwin pattern auger bit
c   Jennings at an auger bit
d   Jennings pattern dowel bit
e   combination bits
f   countersink
g   expansive bit
h   Scotch-eyed auger bit
i   forstner bit
j   turned screw bit
k   flat bit
l   screw sink bits (2)
m   plug cutter

The spinal ratchet screwdriver, featured in
section 5.8.4(c), may be adapted to drill small
short holes and to carry out countersinking
operations.

Gauging the depth of a hole can be done
either by using a proprietary purpose-made

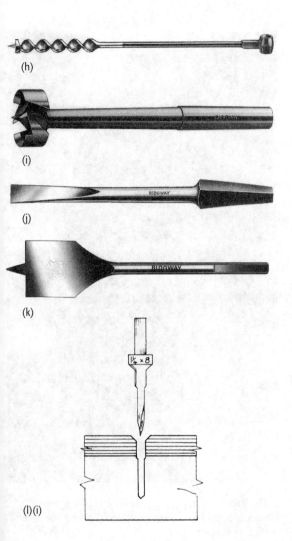

(h)

(i)

(j)

(k)

(l)(i)

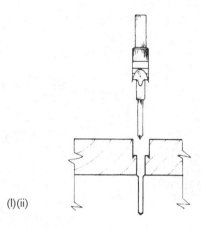

(l)(ii)

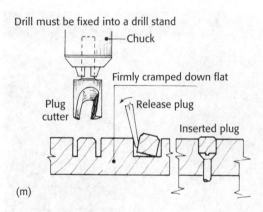

Drill must be fixed into a drill stand

Chuck

Firmly cramped down flat

Plug cutter

Release plug

Inserted plug

(m)

depth gauge or by making your own as in fig. 5.60.

5.6   Chisels (wood)

Note: for cold chisels and bolster chisels see book 3. (Under Repairs & Maintenance)

Woodcutting chisels are designed to meet either general or specific cutting requirements. Table 5.4 lists those chisels in common use, together with their characteristics. Figures 5.61 illustrates some popular examples.

Chisels can be divided into two groups, those that cut by:

- paring, and
- chopping.

5.6.1 Paring process

This simply means the act of cutting thin slices of wood – either across end grain (fig. 5.62) or across the grain's length (fig. 5.63). Chisels used for this purpose are slender and designed for easy handling. The bevel on the edge of a chisel provides for cutting into acute corners.

Figures 5.62 and 5.63 show two examples of paring. **Note** especially the method of support given to the body, Workpiece, and chisel (both hands *always* behind the cutting edge).

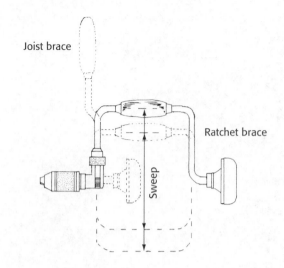

Joist brace

Ratchet brace

Sweep

Fig 5.53 Types of carpenter's brace

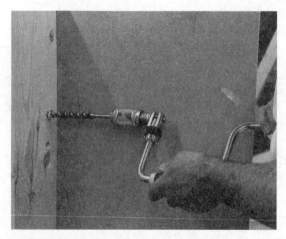

Fig 5.54 Using the ratchet mechanism of a carpenter's brace when full sweep is restricted

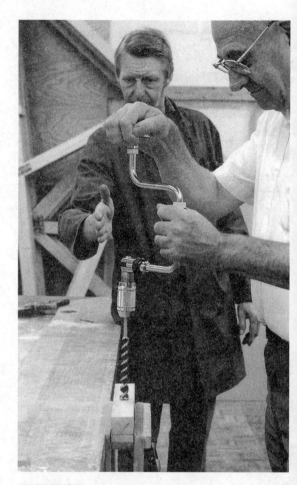

Fig 5.55 Aligning a bit for vertical boring using hand signal guidance from an assistant taking the second vertical viewpoint

5.6.2 Chopping process

Chisels used for this purpose are more robust than paring chisels, to withstand being struck by a mallet. Their main function is to cut (chop) through end grain – usually to form an opening or mortise hole to receive a tenon; hence the common name *mortise chisel*.

Figure 5.64 shows a mortise chisel and mallet being used to chop out a mortise hole.

Fig 5.56 Aligning a bit for horizontal boring using hand signal guidance from an assistant taking the second horizontal viewpoint

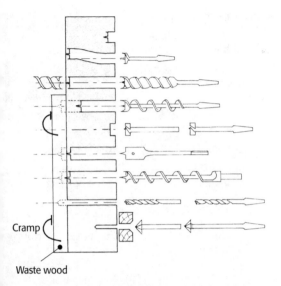

Cramp

Waste wood

Fig 5.57 Application of various bits and drills

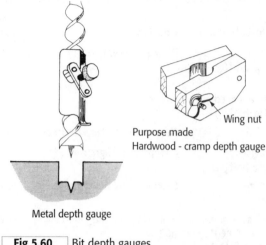

Wing nut

Purpose made
Hardwood - cramp depth gauge

Metal depth gauge

Fig 5.60 Bit depth gauges

Fig 5.58 Hand drill (wheel brace) (with kind permission from Stanley tools limited)

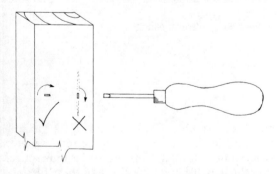

Fig 5.59 The correct method of applying a bradawl

Notice the firm support given to the workpiece, and how the 'L' cramp has been laid over on its side so as not to obscure the operative's vision, and how the protruding bar and handle are positioned towards the well of the bench, out of harm's way.

5.7 Shaping tools

All cutting tools can be regarded as shaping tools – just as different shapes determine the type or types of tools required to form them. The previous tools could therefore also be included under the heading of *shaping tools*.

5.7.1 Axe (blocker)

Provided an axe is used correctly, i.e. always keeping fingers and body behind its cutting edge, it can be a highly efficient tool – invaluable to the site worker for quick removal of waste wood or for cutting wedges etc. It must be kept sharp however, and at the finish of each operation the blade must be protected with a thick leather sheath.

Figure 5.65 shows an axe being used to cut a wedged-shaped plug.

5.7.2 Surform type tools

A very useful, versatile range of shaping tools, capable of tackling most materials, depending on

Table 5.4 Characteristics of woodcutting chisels (not all types are available in metric sizes)

Chisel type	Function	Handle material	Blade widths	Blade section	Remarks
Firmer (Fig. 5.61(a))	Pairing and light chopping	HW plastics	¼–1¼ in 6–32 mm		General bench work etc. Plastics-handle types can be lightly struck
Bevel-edged (Fig. 5.61(b))	As above	HW plastics	⅛–1½ in 3–38 mm		As above, plus ability to cut into acute corners
Pairing	Pairing long or deep trenches	HW	¼–1½ in		Extra-long blade
Registered (Fig. 5.61(c))	Chopping and light mortising	HW	¼–1½ in 6–38 mm		Steel ferrule prevents handle splitting
Mortise (Fig. 5.61(d)	Chopping and heavy mortising	HW plastics	¼–½ in		Designed for heavy impact
Gouge – firmer (out-cannel)* (Fig, 5.61(e))	Hollowing into the wood's surface	HW	¼–1 in 6–25 mm		Size measured across the arc
Gouge – scribing (in-cannel)* (Fig. 5.61(f))	Hollowing an outside surface or edge				Extra-long blades available (paring gouge)

Note: * Out-cannel (firmer) gouges have their cutting bevel ground on the outside; in-cannel (scribing) gouges have it on the inside

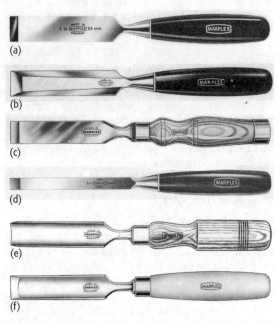

(a)

(b)

(c)

(d)

(e)

(f)

Fig 5.61 Wood cutting chisels
a   firmer chisel
b   bevel edged chisel
c   registered chisel
d   mortice chisel
e   firmer gouge (out – cannel)
f   scribing gouge (in – cannel)

Fig 5.62 Vertical paring

the blade (table 5.5). They can be a valuable asset for joiners involved in house maintenance, where conventional tools are often impractical, and the bench-hand may also find these files and shapers very useful.

Fig 5.63 | Horizontal paring

Fig 5.64 | Chopping a mortice hole – notice the use of scrap wood to protect the bench and work piece, and how the cramp has been laid flat with protruding bar and handle positioned towards the well of the bench, out of harms way

Figure 5.66 shows just two of these tools, and Table 5.5 shows their blade capabilities.

5.8 Driving (Impelling) tools

These tools have been designed to apply a striking or turning force to fixing devices or cutting tools. For example;

- hammers – striking,
- mallets – striking,

Fig 5.65 | Using an axe to cut a taped end on a short length of timber
Notice the piece of timber is positioned on the floor to protect both the floor and the axe cutting edge

- screwdrivers – turning,
- carpenter's braces – turning, | Dealt with under 'boring tools'
- hand drills (wheelbraces) – turning.

5.8.1 Hammers

There are four types of hammer you should become familiar with, these are listed below

Table 5.5 Stanley Surform blades

	Standard cut Plane Planerfile Flat file	Fine cut Plane Planerfile Flat file	Half round Plane Planerfile Flat file	Metals & plastics Plane Planerfile Flat file	Round Round file	Fine cut Block plane Ripping plane	Curved Shaver tool
Hardwoods	■	■	■		■	■	■
Softwoods	■	■	■		■		■
End grain		■				■	
Chipboard	■		■		■	■	
Plywood	■		■		■	■	
Blockboard	■		■		■	■	
Vinyl	■		■		■		
Rubber	■		■		■		
Plaster	■	■	■		■	■	
Thermalite	■	■	■		■	■	
Chalk	■	■	■		■	■	
Glass fibres	■	■	■		■		
Brass		■		■		■	■
Lead		■		■		■	■
Aluminium		■		■		■	■
Copper		■		■		■	■
Mild steel		■		■			
Plastic laminates				■			
Plastic fillers				■		■	■
Nylon	■			■		■	
Linoleum	■				■	■	
Ceramics		■					

together with their imperial and metric head weight.

a claw hammers (16ozs[450gms], 20ozs[570gms], 24ozs[680gms]).

b Warrington or cross-pein hammers (8ozs,No.0.[230gms], 10ozs, No.1.[280gms] 12ozs,No.2.[340gms], 16ozs,No.4.[450gms]).

c engineer's or ball-pein hammers (1lb [450gms], 1½lb [680gms], 2lb [910gms]]

d club or lump hammers (2lb [910gms], 2½lb [1.135 kg], 4lb [1.815 kg]).

a *Claw hammers* (fig. 5.67a) – available with either a shaft of wood, or steel and a handgrip of rubber or leather. They are used

Fig 5.66 Selection of Stanley 'surform' tools (with kind permission from Stanley tools limited)

Claw hammer (a)

Warrington hammer - cross pein (b)

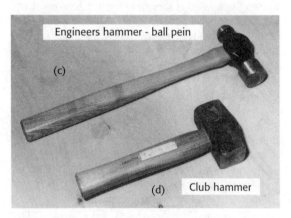

Engineers hammer - ball pein (c)

Club hammer (d)

Fig 5.67 Selection of hammers

for driving medium to large nails and are capable of withdrawing them with the claw. Figure 5.68 shows a claw hammer being used to withdraw a nail – *notice the waste wood used both to protect the workpiece and to increase leverage.*

This type of hammer is the obvious choice for site workers involved with medium to heavy constructional work. It can, however, prove cumbersome as the size of nail decreases.

b *Warrington or cross-pein hammer* (fig. 5.67b) – although capable of driving large nails, this is better suited to the middle to lower range, where its cross pein enables nails to be started more easily. This hammer is noted for its ease of handling and good balance.

Figure 5.69 shows a Warrington hammer being used to demonstrate that, by using the full length of its shaft, less effort is required and greater accuracy is maintained between blows – thus increasing its efficiency. (This applies to all hammers and mallets.)

Tools associated with this hammer are the nail punch (featured in fig. 5.71) and pincers. Pincers (fig 5.83) provide the means to withdraw the smallest of nails.

c *Engineer's or ball-pein hammers* (fig. 5.67c) – the larger sizes are useful as general purpose heavy hammers and can be used in conjunction with wall-plugging chisels etc.

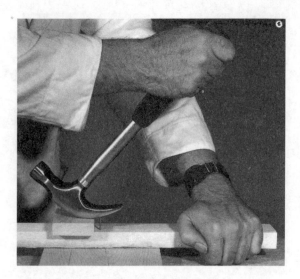

Fig 5.68 Withdrawing a nail using a claw hammer. Leverage should be terminated and a hammer reposition by using a packing (as shown) before the hammer shaft reaches the upright position, or an angle of 90 degrees to the face of the work piece – otherwise undue stress could damage the hammer shaft

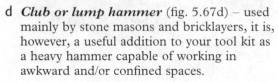

Fig 5.69 Holding a hammer

d **Club or lump hammer** (fig. 5.67d) – used mainly by stone masons and bricklayers, it is, however, a useful addition to your tool kit as a heavy hammer capable of working in awkward and/or confined spaces.

Warning: *Hammer heads should never be struck against one another or any hardened metal surface, as this action could result in the head either splitting or splintering – particles could damage your eyes.*

If a hammer face becomes greasy or sticky, the chances are that your fingers or workpiece will suffer a glancing blow. Always keep the hammer's striking face clean by drawing it across a fine abrasive paper several times.

Wooden shafts are still preferred by many craftsmen, probably because of their light weight and good shock-absorbing qualities. Users of wooden-shafted hammers must however make periodic checks to ensure that the head is secure and that there are no hairline fractures in the shaft. Figure 5.70 illustrates three examples of how a shaft could become damaged.

5.8.2 Punches

We shall consider four types of punch:

a pin and nail punch,
b nail drifter,
c centre punch,
d name punch.

a **Pin and nail punch** (fig 5.71) – struck by a hammer, these punches (depending on the diameter of the cupped head size), sink a pin or nail head just below the wood surface. The resulting small indentation left by the nail head can then be filled with the appropriate filler.

The cupped point helps to seat the punch over the fixing.

b **Nail drifter** – (fig 5.72) – purpose made by the joiner either from an old wall plugging

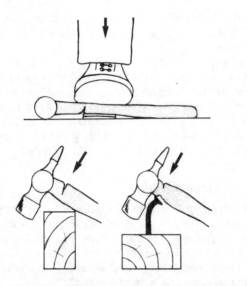

Fig 5.70 | Damage to wooden shaft

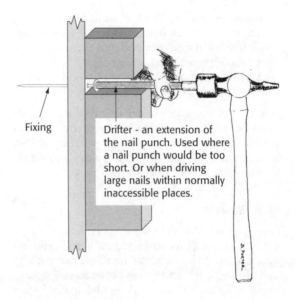

Fixing

Drifter - an extension of the nail punch. Used where a nail punch would be too short. Or when driving large nails within normally inaccessible places.

Fig 5.72 | Nail drifter

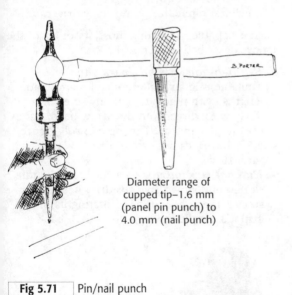

Diameter range of cupped tip–1.6 mm (panel pin punch) to 4.0 mm (nail punch)

Fig 5.71 | Pin/nail punch

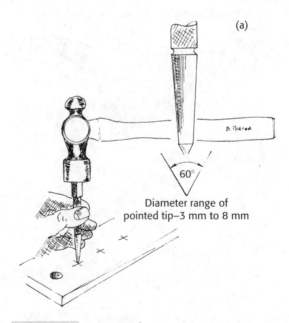

(a)

60°

Diameter range of pointed tip–3 mm to 8 mm

Fig 5.73 | Centre punch

chisel or a short length of 10 mm mild steel rod. The drifter is used as a striking medium between the hammerhead and a round headed nail in inaccessible places where the hammerhead cannot accurately strike the nail head.

c *Centre punch* (fig 5.73) – generally used by the metalworker, but because the joiner is often called upon to drill holes in metalwork,

this punch has been included. It is used to make a small indentation in the metal prior to drilling. Thereby providing the drill bit with an initial pilot and help prevent a drill bit sliding at the start of its first rotation.

d **Name punch** (fig 5.74) – traditionally all joiners tools were personalised by embossing the joiners' name or a code on them as means of recognition and security. Lettered punches are available to stamp wood, plastics and soft metals.

Today, it is more common to mark your tools with your postcode, marks can be made by using special inks or an etching tool – where possible these marks should be concealed and their whereabouts known only to the owner.

5.8.3 Mallets

The head of a mallet, which provides a large striking face, and its shaft, which is self-tightening (tapered from head to handle), are usually made from Beech and weigh between 0.4 kg and 0.6 kg. Choice will depend on the mallet's use and the user. Many joiners prefer to make their own mallet, in which case it can be made to suit their own hand.

The joiner's mallet should be used solely to

| **Fig 5.74** | Name punch |

strike the handle of wood-cutting chisels – figure 5.75 shows its correct use, and figure 5.64 shows it in use.

Using a mallet for knocking together timber frames or joints should be regarded as bad practice, because, unless protection is offered to the surface being struck, the mallet will have a similar bruising effect to a hammer. However, soft headed mallets are available like the one shown in figure 5.76 specifically for this purpose.

5.8.4 Screwdrivers

The type and size of screwdriver used should relate not only to the type and size of screw but also to the speed of application and the location and quality of the work.

There are three basic types of screwdriver used by the carpenter and joiner;
a the fixed or rigid-blade screwdriver,
b the ratchet screwdriver,
c the spiral ratchet or pump screwdriver.

Each is capable of tackling most, if not all, of the screws described in section 12.2

a **Rigid-blade screwdrivers** – these are available in many different styles and blade lengths, with points to suit any screw head. They work directly on the screw head (screw eye) to give positive driving control. Figure 5.77 illustrates two types of rigid blade screwdriver.
b **Ratchet screwdriver** – available to handle slotted and superdriv (posidriv) headed screws and is operated by rotating its firmly gripped handle through 90° – then back –

Tapered shaft

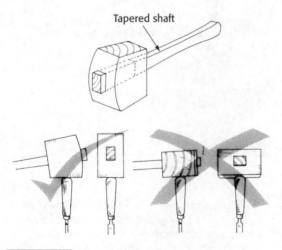

| **Fig 5.75** | Using a wood mallet correctly |

Two examples by size of soft headed mallets

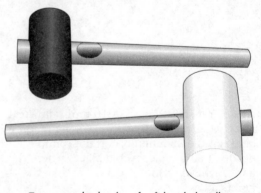

| **Fig 5.76** | Rubber headed soft mallets |

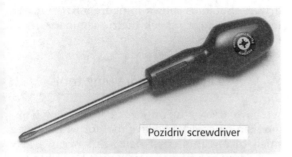

Pozidriv screwdriver

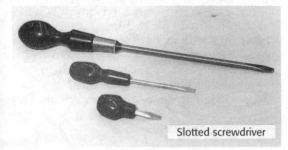

Slotted screwdriver

Fig 5.77 Rigid blade screwdrivers

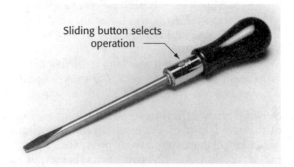

Sliding button selects operation

Fig 5.78 Ratchet screwdriver

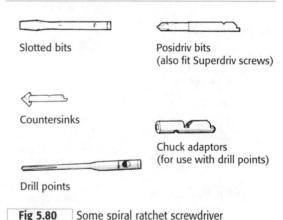

Fig 5.79 Spiral ratchet screwdriver and bits (kind permission of Stanley tools limited)

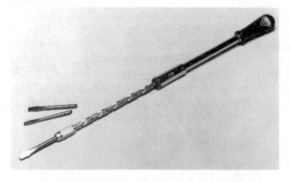

Slotted bits

Posidriv bits
(also fit Superdriv screws)

Countersinks

Chuck adaptors
(for use with drill points)

Drill points

Fig 5.80 Some spiral ratchet screwdriver accessories

and repeating this action for the duration of the screw's drive. A clockwise or counter-clockwise motion will depend on the ratchet setting – a small sliding button, illustrated in figure 5.78, is used to pre-select any of the following three operations;

1 forward position – clockwise motion,
2 central position – rigid blade,
3 backward position – counter-clockwise motion.

Because the driving hand always retains its grip on the screwdriver throughout its operation, this ratchet facility speeds up the process.

c *Spiral ratchet screwdriver* – this is often termed a pump screwdriver, because of its pump action, and is by far the quickest hand method of driving screws. Not only can it handle all types of screws, it can also be adapted to drill and countersink holes.

   Figure 5.79 shows a Stanley *Yankee* spiral ratchet screwdriver, and figure 5.80 indicates some of the many accessories available. Its ratchet control mechanism is similar to that of the standard ratchet screwdriver. However, its driving action is produced by pushing (compressing) its spring-loaded barrel over a spiral drive shaft, thus rotating the chuck (bit-holder) every time this action is repeated.

Warning: by turning its knurled locking collar, the spiral drive shaft can be fully retained in the barrel, enabling it to be used as a short rigid or ratchet screwdriver. But, while the shaft is spring-loaded in this position, its point must always be directed away from the operator, as it is possible for the locking device to become disengaged, in which case the shaft will lunge forward at an alarming rate and could result in serious damage or injury.

After use, always leave this screwdriver with its spiral shaft fully extended. The spring should never be left in compression.

Screwdriver efficiency – with the exception of the *stub* (short-blade) screwdriver, the length of blade will correspond to its point size. If driving is to be both effective and efficient, it is therefore important that the point (blade) must fit the screw eye correctly.

Figure 5.81(a) illustrates how face contact with a slotted eye can affect the driving efficiency – (i) and (ii) are inclined to come out of the slot; therefore (iii) should be maintained at all times. Similarly, if the blade end is too thin or rounded at the corners it will damage the screw as it is rotated within the screw eye (also see fig 5.125).

Point style and size of cross-headed screw-

drivers should as shown in figure 5.81(b) match the screw gauge (check metric comparisons in tables 12.3 & 12.4).

5.9 Lever & withdrawing tools

We have already mentioned that the claw hammer can be a very effective tool for withdrawing small to medium-sized nails. But where heavy application is required using a wrecking bar (fig 5.82(a)) can provide extra leverage – it is sometimes termed a nail bar or Tommy bar. A smaller version for medium duty leverage is shown in figure 5.82(b).

Alternatively as shown in figure 5.83 when withdrawing small nails or panel pins a pair of pincers can be used. The jaws of the pincers are very

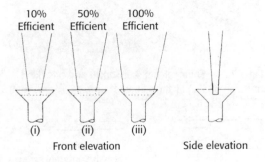

10% Efficient   50% Efficient   100% Efficient

(i)   (ii)   (iii)

Front elevation   Side elevation

a) Slot-headed screws

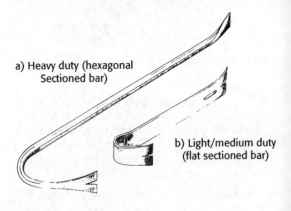

a) Heavy duty (hexagonal Sectioned bar)

b) Light/medium duty (flat sectioned bar)

Fig 5.82   Wrecking bars

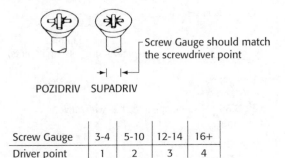

Screw Gauge should match the screwdriver point

POZIDRIV   SUPADRIV

Screw Gauge	3-4	5-10	12-14	16+
Driver point	1	2	3	4

N.B. No.2 is the most useful size

b) Cross-headed screws

Fig 5.81   Guide to screwdriver efficiency

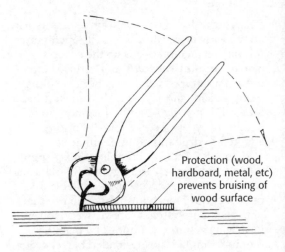

Protection (wood, hardboard, metal, etc) prevents bruising of wood surface

Fig 5.83   Pincers – withdrawing a small nail

strong and therefore capable of cutting through soft metal and small sections of mild steel wire, this in many cases can be an added bonus.

5.10   Finishing tools & abrasives

The final cleaning-up process will be determined by whether the grain of the wood is to receive a transparent protective coating or is to be obscured by paint. In the former case, treatment will depend on the type and species of wood, whereas the treatment for painting is common to most woods.

Hardwood – because it's main use is decorative, hardwood is usually given a transparent protective coating. Unfortunately, the grain pattern of many hardwoods makes them difficult to work – planing often results in torn or ragged grain. A scraper can resolve this problem, and should be used before finally rubbing down the surface with abrasive paper.

Note: *always follow the direction of the grain when using abrasive paper before applying a transparent finish (see fig. 5.87).*

Softwood – softwood surfaces that require protection are usually painted – there are, however, some exceptions. In preparing a surface for paint, flat surfaces must be flat but not necessarily smooth – unlike transparent surface treatments, minor grain blemishes etc. will not show. Surface ripples caused by planing machines do show, however, and will require levelling with a smoothing plane (alternatively mechanical sanding methods can be used), after which an abrasive paper should be used. The small scratch marks left by the abrasive help to form a key between the wood and its priming paint (first sealing coat).

5.10.1 Scrapers

These are pieces of hardened steel sheet, the edges of which (as shown in figure 5.84(a)) have been turned to form a burr (cutting edge).

A scraper is either of a type which is held and worked by hand or is set into a scraper plane which locks and is held, like a large spokeshave. Scrapers are slightly bent during use, to avoid digging their sharp corners into the wood (Fig. 5.84(b)). Scraper planes, however, can be pre-set to the required cutting angle and then simply require pushing, or dragging as the case may be.

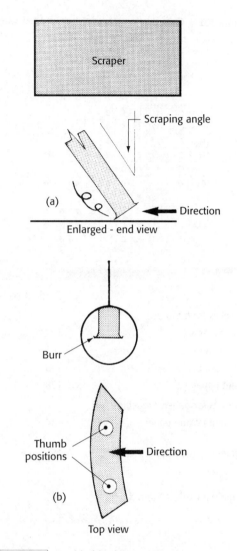

Fig 5.84 Hand held cabinet scrape

5.10.2 Sanding

The application of abrasive-coated paper or cloth to the surface of wood.

There are several kinds of grit used in the manufacture of abrasive sheets, the three most popular being glass, garnet, and aluminium oxide. They are available in a sheet size 280 mm × 230 mm and are graded according to their grit size, which ultimately determines the smoothness of the wood.

Table 5.6   Comparative grading of abrasive sheets

Grain	Backing	Product code	Bond	Coat	Sheets	Coils	Rolls	Belts	Discs	1200	1000	800	600	500	400	360	320	280	240	220	180	150	120	100	80	60	50	40	36	30	24	16
Glass	Paper C	117	G	C	●														FL	0	1	1½	F2		M2*		S2	2½	3			
	Paper C	121	G	C	●														00	0	1	1½	F2		M2*		S2	2½	3			
Garnet	Paper A	129	G	O	●											9/0			7/0	6/0	5/0	4/0	3/0	2/0	0							
	Paper C + D	128	G	O	●																5/0	4/0	3/0	2/0	0	½	1	1½				
Aluminium oxide	Paper C	134	G	O	●																●	●	●	●	●	●	●	●	●			
	Paper E	156	R/G	C	●	●	●	●							●			●	●	●	●	●	●	●	●	●	●	●	●	●	●	●
	Paper E	157	R	C				●	●										●	●	●	●	●	●	●	●	●					
	Cloth X	141	R	O	●	●	●	●	●										●	●	●	●	●	●	●	●	●	●				
	Cloth X	147	R	C			●	●	●									●	●	●	●	●	●	●	●	●	●	●			●	
	Cloth FJ	144	R	C				●								●	●	●	●	●	●	●	●	●	●	●	●				
	Cloth J	145	R	C					●									●	●	●	●	●	●	●	●	●					

Note: Bond: Glue – G; Resin – R; Resin/Glue – R/G; Grain configuration: Open coat – O; Closed coat – C; Grain size: M2 = 70 grit

	Product code	Description
Abrasives for hand use		
Glass paper – C	117	Cabinet quality for joinery and general wood preparation
Glass paper – C	121	Industrial quality for joinery and general wood preparation
Garnet paper – A	129	For fine finishing and contour work on furniture
Garnet paper – C 1 D	128	Durable product for fine finish stages of wood preparation
Abrasives for general machine use		
Aluminium oxide paper – C	134	For rapid stock removal with reduced clogging. For hand and orbital sander use
Aluminium oxide paper – E	156	For sanding medium hardwoods and softwoods. A durable product recommended for heavy hand sanding, orbital sanders, and pad and wide belt sanders
Aluminium oxide – Cloth X	141	General belt sanding of wood, particularly recommended for portable sanders, open coat configuration reduces clogging

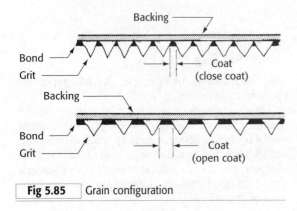

Fig 5.85   Grain configuration

Some comparative gradings are listed in table 5.6, whilst figure 5.85 should help explain the differences in grain configuration between 'open coat' and 'closed coat', and table 5.7 which should help you to choose the right coating in relation to the abrasives application.

By using tables 5.6 & 5.7 together with figure 5.86 you should be able to unravel the codings used on the backs of abrasive sheets.

A sanding block should always be used to ensure a uniform surface, be it flat, curved or round. Typical examples are shown in figure 5.87, together with a method of dividing a sand-

Table 5.7   Choosing the correct coated abrasive for the type of work

Key	Application
Backing	
Paper	A lower cost material for less demanding uses
Cloth	Higher tensile strength, strong and durable
Bond	
(G) All glue	Used for hand and light machine applications, a flexible bond
(R/G) Resin over Glue	Intermediate flexibility and durability
(R) All resin	Higher band strength and heat resistance, for more demanding applications
Grain Configuration	
(O) Open coat	Reduced clogging, coarser finish
(C) Closed coat	Finer finish, faster cutting

Grit size
The coarser the grit (larger particles, the smaller the grit size number, according to international FEPA 'P' standards. The non-technical abrasives are graded according to different standards and are equivalent as shown in the table.

Note: grit (particle) size – the smaller the grit number, the more coarse the grit. For example, grit number 24 is very coarse and grit numbers 50–70 medium, while grit number 400 is very fine

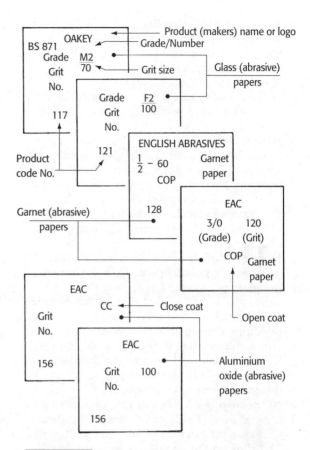

Fig 5.86   Coding on the back of some coated abrasive sheets

ing sheet into the appropriate number of pieces to suit the blocks.

5.11   Holding equipment (tools & devices) – Also see Practical Projects No. 6

Holding tools are general-purpose mechanical aids used in the preparation and assembly of timber components. Holding devices have usually been contrived to meet the needs of a specific job or process.

5.11.1   Holding tools

Several examples are listed below together with a brief description:

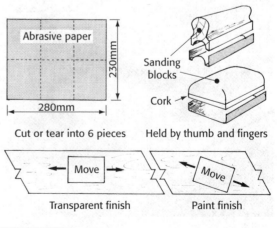

Fig 5.87   Sanding by hand

- **_Joiner's bench vice_** (fig 5.96(a) & Project 6 fig 24) – large hardwood faced wide opening jaws, single turn screw operation with two solid steel bar guides and a quick release opening and closing mechanism.

  Some models have a height adjustable steel 'dog' set into the face of the front jaw – used in conjunction with 'bench dogs'.

- **_Bench Dogs_** (Project 6 fig 24) – square, or round (with one flat face) pegs with integral springs to fit into holes cut into the bench worktop. They work in conjunction with the bench vice dog to produce a bench top cramping facility for holding frames, or short lengths of timber whilst being worked on.

- **Bench holdfast** (figs 5.88 & 5.96) – there are several types all operating in similar way. Some simply fit into a hole bored into the bench top. Others, like the one illustrated, fit into a metal collared tube. They work on the principle of using an offset counter-lever, the arm is free to pivot over the bar offset, until leverage forces instigated by the downward movement of the turn-screw, move it downwards onto the workpiece.

- **_'G' cramp_** (fig 5.89) – as the name implies it is 'G' shaped – different sizes are available with cramping depths of 50 mm to 300 mm. The shoe on the end of the threaded spindle swivels on a ball socket enabling it to remain still whilst the turn screw is being tightened.

- **_Joiner's hand screw ('F' cramp)_** (fig 5.90) -although not as strong as a 'G' cramp, they are often favoured for lighter work because of the effective quick clamping and release mechanism.

- **_Trigger cramp_** (single handed clamp – fig 5.91) – forward moving bar within a 'C' frame activated by squeezing a trigger – pressure exerted limited to the grip of the operative.

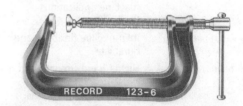

Fig 5.89   'G' cramp

Fig 5.88   Bench hold fast

Fig 5.90   'F' cramp

Sash cramp (fig 5.92) – consists of two cramp shoes, one (the head shoe) is fixed to a screw spindle to provide adjustment over a limited length. The second shoe (tail shoe) can slide over the remaining length of the rectangular steel bar to a pre-determined length, any further movement to facilitate the cramping process being restricted by the insertion of a steel pin into any one of the holes in the bar.

'T' bar cramp (fig 5.93) – works on the same principle as a sash cramp but in the main due to the 'T' profile of the bar it is much stronger. Lengthening 'T' bars are also available for those large jobs.

Cramping heads (fig 5.94) – a light duty alternative to the sash cramp. The bar in this case is made-up of a length of 25 mm wide timber. Holes to receive the pins are bored as and where required along the timber bar.

Dowelling jigs (fig 5.95) – these are used to ensure that holes drilled into two eventual adjoining members are accurately positioned to receive dowels. These jigs usually have provision for dowel diameters of 6 mm, 8 mm, & 10 mm.

5.11.2 Holding devices

Several examples are listed below together with a brief description:

Bench stops (fig 5.96 & Project 6 figs 22 & 23) – square hardwood peg mortised into the bench top to provide an end-stop whilst planing faces of timber. Proprietary bench stops (Project 6 fig 23) consist of a metal

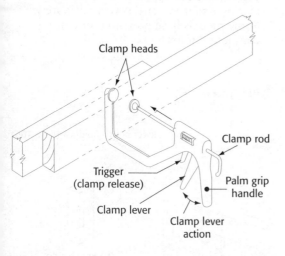

Fig 5.91   Single-handed lever trigger cramp (ratchet type)

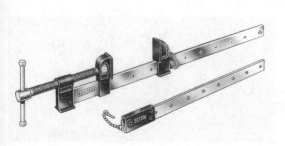

Fig 5.92   Sash cramp

Fig 5.93   'T' bar cramp

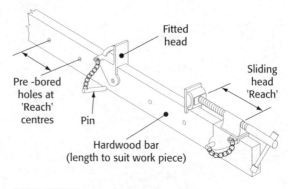

Fig 5.94   Detachable cramp heads

Fig 5.95 Dowelling jig (with kind permission from Neil tools)

boxed plate – one end hinged, the other serrated (end stop), can be raised (up 12 mm) above the bench surface when in use, or lowered flush with the bench top out of harms way when not required.

- *Bench hook* (fig 5.96(i)) – bench top hand holding device for cross cutting short lengths of timber. Also see figure 5.17(b).
- *Mitre block* (fig 5.96(j)) – a device used to support squared or moulded sections, usually held secure within the bench vice or 'workmate' jaws whilst making cuts at an angle (usually 45°) across small sectioned timber or beads in conjunction with either a tenon or dovetail saw.
- *Mitre box* (fig 5.96(k)) – same function as a mitre block, but potentially offering greater accuracy – usually deeper or wider sections of timber.
- *Mitre box and saw* (fig 5.97) – a proprietary combination of saw and jig for cutting a range of angles very accurately. Some models have provision for cutting compound mitres.

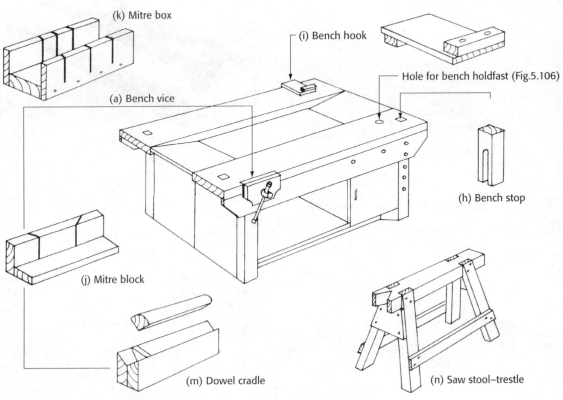

(k) Mitre box

(i) Bench hook

Hole for bench holdfast (Fig.5.106)

(a) Bench vice

(h) Bench stop

(j) Mitre block

(m) Dowel cradle

(n) Saw stool–trestle

Fig 5.96 Joiner's work bench & holding equipment (also see projects 6 & 7)

Table 5.8 Holding equipment (see Figure 5.96)

Equipment (see Fig. 5.96)	Use	
	Preparing material	Assembly aid
Holding tool		
a) Bench vice	Yes	Some models
b) Bench holdfast Figs. 5.88)	Yes	No
c) 'G' cramp (Fig. 5.89)	Yes	Yes
d) 'F' cramp (Fig. 5.90)	Yes	Yes
e) Sash cramp (Fig. 5.92)	Yes	Yes
f) 'T' bar cramp (Fig. 5.93)	No	Yes
g) Dowelling jig (Fig. 5.95)	Yes	
Holding devices		
h) Bench stop*	Yes	No
i) Bench hook*	Yes	No
j) Mitre block	Yes	No
k) Mitre box	Yes	No
l) Mitre box and saw (Fig. 5.97)	Yes	No
m) Dowel cradle	Yes	No
n) Saw stool (trestle) See project 7	Yes	Yes
o) Stanley workmate (Fig. 5.98)	Yes	Yes

Note: * Provision can be made in their construction for left- or right-handed users

- ***Dowel cradle*** (fig 5.96(m)) – a vice held cradle, designed to hold right-angled timber sections whilst planing corners round.
- ***Saw stool, or Trestle*** (fig 5.96(n) also see project 7) – usually used in pairs, either to form a low bench or for support while sawing.
- ***Stanley workmate*** (fig 5.98) – a strong folding trestle. The top is in two parts which serve as a vice and /or cramp capable of holding the most awkward of shapes.
- ***Wolfcraft workbench*** (fig 5.99) – similar function to the 'Workmate' but with an option of hands free clamping and releasing. As shown these hands free operations are achieved via a double – single foot operated split pedal. To close the jaws with a parallel motion the pedal is pressed centrally – to alternate jaw closing (right or left) the appropriate pedal side is pressed.

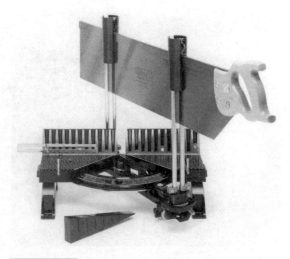

Fig 5.97 Box and mitre saw

5.12 Tool storage & accessory containers

The variety and condition of the tools used by the craftsperson often reflects the quality and type of work capable of being undertaken by them. Apart from your basic tool kit of, say,

a four-fold metre rule,
- flexible steel tape,
- combination square,
- claw hammer,
- handsaws (traditional panel saw, and/or hardpoint), and possibly a Tenon saw,
- jack and smoothing planes,
- assortment of chisels – firmer/bevel-edged,
- carpenter's brace,
- assortment of twistbits and countersink,
- hand drill (wheel-brace) – generally superseded by the cordless drill,
- assortment of twistdrills,
- *screwdrivers,
- bradawl.

(**Note:** sharpening accessories (oilstones etc.) should also be included on this kit.)

Now commonplace to rely on the cordless drill and screwdriver, in which case these hand-tools should still be regarded as essential backup tools.

There is no doubt that irrespective of the kind of work you are employed to do, over the forthcoming years, many different types of tool will be acquired – all of which will require some form of protection and safe storage provision – whether you are to be site or workshop based.

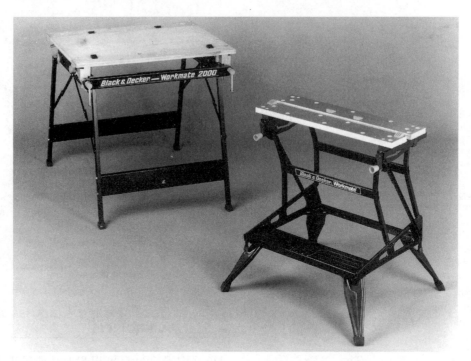

Fig 5.98 Stanley workmate

Enlarging the immediate number and type of tools in your kit will depend on your job specification. Broadly speaking, tool kits can be broken down into three groups (see tables 5.1–5):

1 *Everyday use* – basic tool kit.
2 *Occasional use* – the more common of the special use tools.
3 *Specific use* – special use tools.

In time you will find that you are duplicating some of your everyday tool kit. For example; having at least two hand saws should ensure that one is always kept sharp.

No matter what means of storage is chosen, the most important factors to consider about a container, are that each item is housed separately or is individually protected from being knocked against another, and that all cutting edges are enclosed or sheathed in some way. Of course the body of the tool must also be protected from damage – unfortunately many joiners seem to regard this form of protection as secondary.

Always bear in mind that a well-worn tool is not one sporting the battle scars of its container but which, after many years of active service, appears virtually unscathed.

There are a variety of ways by which you may

want to house your tools – the method usually chosen often reflects the type of work your company undertakes. The final choice is usually yours. Different methods of containing tools include;

- traditional toolchest (fig. 5.100(b))
- traditional toolbox (fig. 5.100(a))
- Porterbox (fig 5.100(c) & 5.101)
- tool case (fig. 5.100(d))
- tool tray (usually racked) (fig. 5.100(e))
- tool bag (fig. 5.100(f))
- joiner's bass (fig. 5.100(g))

A combination of the above is usually chosen.

5.12.1 The traditional toolchest (fig. 5.100(b))

A very strong, secure, top opening rectangular box of solid construction. It is designed to accommodate all of the joiners' tools (everyday and special usage). Larger tools are housed within base compartments, whereas smaller tools are separated within a series of compartmentalised sliding lift-out trays. The two upper trays being half the box length, thereby allowing access to box contents without their removal.

Transporting such a chest is a two-man oper-

Fig 5.99 | Wolfcraft workbench

ation, therefore heavy-duty drop-down chest handles bolted one to each end are essential.

5.12.2 The traditional toolbox (fig. 5.100(a))

A strong rectangular drop fronted box long enough to house the longest handsaw – generally suitable for both workshop and site use. Top, bottom and sides of timber are dovetail jointed at the corners. Plywood is glued, screwed and/or nailed to this framework to form the front and back. Once set, a front portion is cut out to form a hinged flap – this flap houses the handsaws. The interior can accommodate one or more drawers.

I have always regarded this arrangement as unsatisfactory for a number of reasons, the two main ones being:

- *In order to gain access to the tools the flap has to take up a lot of floor space, and if left open is a tripping hazard to passers by.*
- *Because the top of the box is often used as a saw stool, and seat during break periods, the handle and closing/security mechanism is in the way.*

It was with these factors in mind that I designed and developed a system of boxes known as the *Porterbox System* which consists of a:

- Toolbox (Porterbox) (fig 5.101)
- Accessories chest (Porterchest) (fig 5.104)
- Tool case (fig 5.101(c)
- Tool tray (fig 5.101(b)

These are featured at the end of this book as work projects.

5.12.3 Porterbox (fig 5.100(c) & 5.101(a))

An easily constructed box which is strong, secure, and stable. It not only safely houses all your basic tools, with easy access, but can also serve as a small work bench and saw stool. Its working height can be increased or decreased by the add-on Portertray (fig 5.101(b)) or Portercase (fig 5.101(c)).

Its portability can be enhanced by the addition of the Portercaddy (fig 5.101(d)).

5.12.4 Tool case (fig. 5.100(d))

There are occasions when the intended job does not warrant taking all your basic toolkit. On these occasions a smaller, lighter box (scaled down version of your traditional toolbox) may be the answer. In this case the length of the saw may not be the main criterion in its design, because a separate saw case (see fig, 5.102(a)) could be used along with the tool case. However, a tenon saw should be capable of being fitted within the lid of the tool case.

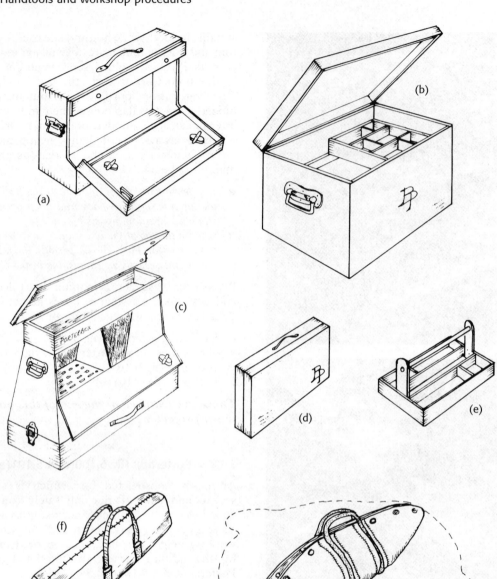

Fig 5.100   Methods of containing and carrying tools and a small ancillary equipment

5.12.5 **Tool tray** (fig. 5.100(e))

An open top box with a box length carrying handle, used as a means of transporting your tools from job to job within the confines of the work area (site). It may also serve as a means of keeping tools tidy when working with other trades, a rack for chisels etc. is a useful feature.

After each work period tools are then trans-ferred back to their secure main tool box, or chest.

5.12.6 **Tool bag** (fig. 5.100(f))

An elongated bag of hard wearing material long enough to contain a handsaw, and strong enough to hold a variety of tools. It's great advantage

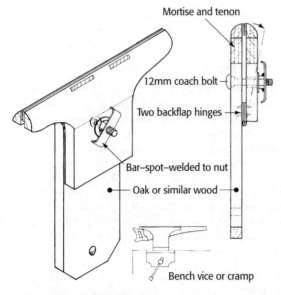

Fig 5.109   Saw vice

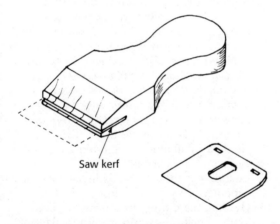

Fig 5.110   Spokeshave blade holder

5.13.2 Saws

Of all tools, saws are the most difficult to sharpen. The problem can be lessened if sharpening is carried out at frequent intervals, because as the condition of the saw worsens, so does the task of re-sharpening. Signs of dullness (bluntness) are when the teeth tips becoming shiny or extra pressure being needed during sawing. The sharpening technique will vary with the type of saw.

If teeth become very distorted, due to inaccurate sharpening or accidentally sawing nails etc.,

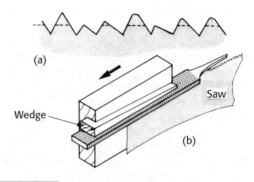

Fig 5.111   Topping a saw blade

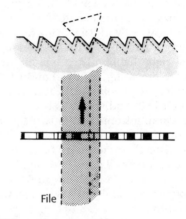

Fig 5.112   Shaping saw teeth

then the following processes should be undertaken;

- topping,
- shaping,
- setting,
- sharpening.

a **Topping** – bringing all the teeth points in line (fig. 5.111(a)) by lightly drawing a long mill saw file over them. A suitable holding device is shown in figure 5.111(b).

b **Shaping** – restoring the teeth to their original shape and size. The file face should be just more than twice the depth of the teeth and held level and square to the saw blade throughout the process (see fig. 5.112). Shaping is the most difficult process to perfect.

c **Setting** – bending over the tips of the teeth to give the saw blade clearance in its kerf (see fig. 5.23).

Manufacturers and saw *doctors* may use a special cross-pein hammer and a saw anvil for this purpose. A suitable alternative is to use saw set pliers, which can be operated with one hand simply by pre-selecting the required points per 25 mm of blade, placing over each alternate tooth, and squeezing (fig. 5.113).

d **Sharpening** – putting the cutting edge on the teeth. Cramp the saw, with its handle to your front, as low as practicable in its vice. Saw teeth of the over hanging part of the saw should be covered (use saw blade sheath – fig. 2.102(b)). Place your file against the front face of the first 'toe' tooth set towards you and the back edge of the tooth facing away from you, then angle the file to suit the type of saw or work (fig. 5.114(a)) and, keeping it flat (fig. 5.114(b)) working from left to right, lightly file each alternate V two or three times (fig. 5.114(c) and (d)). After reaching the handle (heel), turn the saw through 180° and repeat as before, only this time working from right to left (fig. 5.114(c) and (e)).

If the teeth are in a very bad condition, you would be well advised to send your saw to a saw doctor (a specialist in saw maintenance) until you have mastered all the above processes.

Fig 5.113 Setting saw teeth. **Note:** exposed teeth should be covered with a blade sheath (Fig. 102b & c)

Note: *whenever a saw is not in use, ensure that its teeth are protected.*

5.13.3 **Planes**

Plane *irons* (blades) require a grinding angle of 25° and a *honing* (sharpening) angle of 30° (fig. 5.115) if they are to function efficiently – with one exception: Plough planes use a common 35° grinding and honing angle.

The blade shape should correspond to one of those shown in fig. 5.116. A smoothing-plane blade has its corners rounded to prevent digging in, whereas a jack-plane blade is slightly rounded to encourage quick easy removal of wood. A rebate-plane blade must be square to the iron, for obvious reasons.

a **Honing** (sharpening using an oilstone (fig. 5.117)) – this requires much care and attention. The following stages should be carefully studied.

1 Remove the cap iron from the cutting iron – use a large screwdriver to remove the cap screw, *not* the lever cap, otherwise its chrome coating will soon start to peel.

   Note: *this operation should only be carried out whilst the cap iron and cutting iron are laid flat on the bench with some provision made to prevent the cutting edge rotating dangerously – for example, held within and up against the side of the bench well.*

2 Apply a small amount of honing oil to the surface of the oilstone.

3 Ensure that the oilstone is firmly held or supported.

4 Holding the cutting iron firmly in both hands, position its grinding angle flat on the stone, then lift it slightly – aim at 5°, (fig. 5.117(a)).

5 Slowly move the blade forwards and backwards without rocking (fig. 5.117(b)) until a small burr has formed at the back. It is important that the oilstone's entire surface is covered by this action, to avoid hollowing if the blade is narrower than the oilstone (fig. 5.117(c)).

6 The burr is removed by holding the iron perfectly flat, then pushing its blade over the oilstone two or three times (fig. 5.117(d)).

7 The wire edge left by the burr can be removed by pushing the blade across the

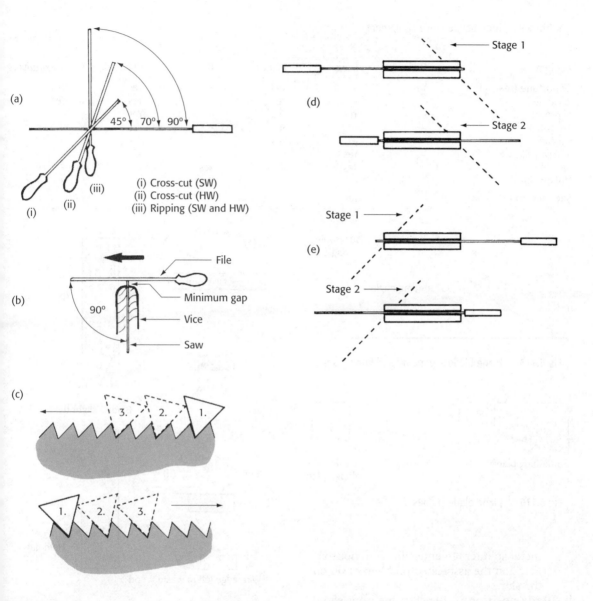

(a)

45° 70° 90°

(i) Cross-cut (SW)
(ii) Cross-cut (HW)
(iii) Ripping (SW and HW)

(i) (ii) (iii)

(b)

File

90°

Minimum gap

Vice

Saw

(c)

3. 2. 1.

1. 2. 3.

(d)

Stage 1

Stage 2

(e)

Stage 1

Stage 2

Fig 5.114 Sharpening process for saw teeth. **Note:** the overhanging portion of the saw blade should be covered with a blade sheath to protect the operative from the sharp teeth

corner of a piece of waste wood (fig. 5.117(e)).

8 If a white line (dullness) is visible on the sharpened edge, repeat stages 5–7. If not, proceed to stage 9.

9 Holding the iron as if it were on the

oilstone, draw it over the strop stick (fig. 5.117(f).

Warning – The practice of showing off by using the palm of the hand as a strop is dangerous and silly – it can not only result in a cut hand or wrist, but may also cause

Table 5.9	Tool maintenance equipment					
	Tool					
Equipment	Drills	Saws	Planes	Bits	Chisels	Screwdrivers
Grindstone (machine)	Yes	–	Yes	–	Yes	Yes
Oilstone	–	–	Yes	–	Yes	Yes
Slipstone	–	Yes	Yes	–	Yes	–
Stropstick	–	–	Yes	–	Yes	–
Oil can	–	–	Yes	–	Yes	–
Mill saw file	–	Yes	–	–	–	–
Saw file	–	Yes	–	Yes	–	–
Needle file	–	–	–	Yes	–	–
Saw vice-clamp	–	Yes	–	–		–

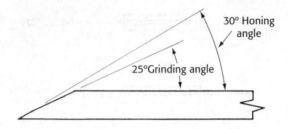

Fig 5.115   Plane blade – grinding and honing angles

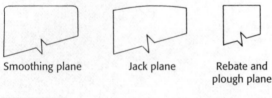

Smoothing plane   Jack plane   Rebate and plough plane

Fig 5.116   Plane blade shapes

metal splinters to enter the skin, not to mention the associated problem of oil on the skin.

b **Blade positions** – Bench-plane irons should be positioned as illustrated in figure 5.118 to ensure effective cutting. The position and angles of *cutting* irons without cap irons can vary, as can be seen in figure 5.119, which shows the arrangement for rebate, plough, and block planes.

5.13.4 Chisels

a **Flat-faced chisels** – these should be ground and honed to suit the wood they are to cut, as shown in figure 5.120. However, where access to a grinding machine is

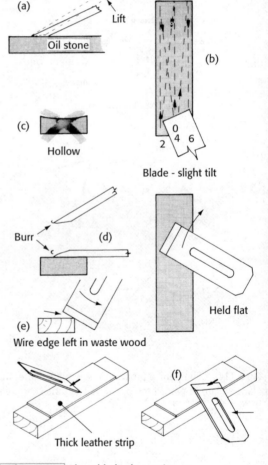

Fig 5.117   Plane blade sharpening process

difficult, i.e. on-site work, it is often possible to extend the useful life of the grinding angle as shown in figure 5.121.

The same principles of honing apply as for

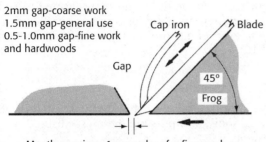

2mm gap-coarse work
1.5mm gap-general use
0.5-1.0mm gap-fine work
and hardwoods

Gap

Cap iron   Blade

45°
Frog

Mouth opening - 1mm or less for fine work or
hardwoods. 2mm for general use

Fig 5.118 Plane back irons and mouth opening
adjustments

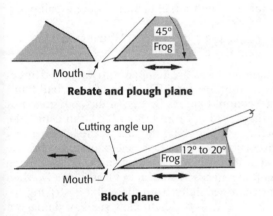

45°
Frog

Mouth

Rebate and plough plane

Cutting angle up

12° to 20°
Frog

Mouth

Block plane

Fig 5.119 Rebate, plough and block plane blade
arrangement

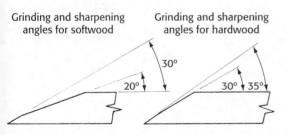

Grinding and sharpening
angles for softwood

Grinding and sharpening
angles for hardwood

30°

20°

30° 35°

Fig 5.120 Chisel grinding and the honing angles

plane irons, but because of the possible
narrowness of the chisel blade extra care
must be taken not to hollow the centre
portion of the oilstone.

b **Gouges** – Firmer gouges are ground on a

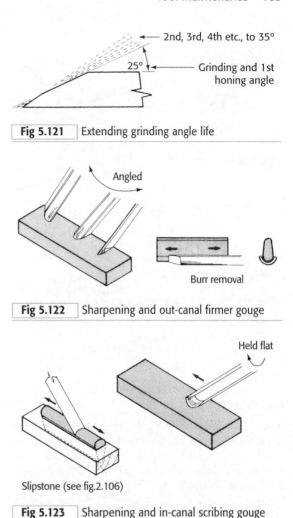

2nd, 3rd, 4th etc., to 35°

25°   Grinding and 1st
honing angle

Fig 5.121 Extending grinding angle life

Angled

Burr removal

Fig 5.122 Sharpening and out-canal firmer gouge

Held flat

Slipstone (see fig.2.106)

Fig 5.123 Sharpening and in-canal scribing gouge

conventional grinding machine – slowly
rocking the blade over its abrasive surface.
Figure 5.122 shows the method of honing
and burr removal. Scribing gouges are
ground on a special shaped grinder and
honed on a slip stone as shown in figure
5.123.

5.13.5 Bits

A twist bit's life is considerably shortened every
time it is sharpened, so be sure that sharpening
is necessary. The spurs (wings) are usually the
first to become dull, followed by the cutters. If
the screw point becomes damaged, replace the
bit. Figure 5.124 shows the method of sharpening a twist bit.

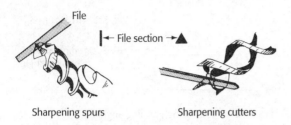

File

|← File section →▲

Sharpening spurs　　　　Sharpening cutters

Fig 5.124 | Sharpening a twist bit

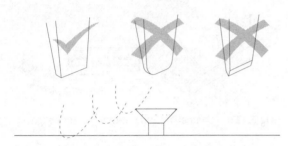

Fig 5.125 | Screwdrivers for slotted screws

5.13.6 Screwdrivers

These are probably the most misused of all hand tools. They have been known to be used as levers, chisels, and scrapers. This is not only bad practice; it is also dangerous. Points do, however, become misshapen after long service, even with correct use. Figure 5.125 illustrates how a point can become misshapen, and the possible consequences. The point should be re-formed to suit the screw eye by filing (using a fine-cut file), rubbing over an oilstone, or regrinding (see section 9.10).

5.13.7 Hammers

Next to the saw probably the most used of all hand tools, and possibly the most misused. For example:

- Never use a hammer with a damaged head.
- Never use a hammer with a loose head.
- Never strike two hammer heads together.
- Never use the side of a hammer.
- Never strike a cold chisel or hardened steel objects with a nail hammer – use a ball pein hammer.
- Never use a cross-pein (Warrington) hammer as a lever.

Hammer faces are tempered to varying degrees depending on type. The face of your nail hammers should always be free of dirt, oil, grease or resin picked up from timber. This can usually be achieved by rubbing the face of the hammer in a circular motion over a sheet of fine abrasive paper that is held over a flat surface. Signs that a hammer needs this treatment are usually when the hammer slides from the nail it is striking.

Should a wood shaft show signs of damage it should be replaced. A fractured shaft or loose head can be very dangerous.

Portable Electric Mains Powered Hand Tools & Machines

6

This chapter deals with portable powered hand tools associated with the motive power supplied by mains electricity and electricity derived from portable generating sets. Other motive powers are obtained from:

- batteries (cordless tools), see chapter 7.
- compressed air (pneumatic tools), (not featured in this book)
- explosive power (cartridge-operated fixing tools), see chapter 8.

6.1 Specification plate (SP)

A label like the one shown in figure 6.1 is fixed to the outer casing of all portable power tools, giving information under some, if not all, of the following headings:

a *Manufacturer – maker's* name or trade mark.
b *Type number – provides* a method of identification for attachments or spare parts.
c *Capacity – chuck* opening size, spindle speed: revolutions per minute (rev/min or RPM) for 'high' and 'low' speed ratings
d *Electrical information*
- *Voltage (V)* -potential difference (electrical pressure) – the supply must be within the range stated on the SP for example 220/224 V or 105/130 V.
- *Hertz (Hz)* – unit of frequency. Indicates the number of cycles per second
- *Amperage (amp)* – the current (rate of flow)taken by the motor, the strength of the electric current expressed in amperes. This can be related to power and voltage

$$\text{Amperes} = \frac{\text{Watts}}{\text{Volts}}$$

- *Wattage* (input on full load) describes the power, or rate of electricity consumption by the motor, expressed in watts, as Watts = Volts × Amperes. Therefore if the potential difference (voltage) is reduced from 240V to 110V as with low voltage tools, the rate of flow (amperage) must be increased if the power output is to be retained. This is why a heavier cable is used with low voltage (i.e. 110V) tools to provide for the increase in current flow.

Watts = Volts × Amperes
Example   1220 W = 240V × **5 amp**
                 1220 W = 110V × **11 amp**

e Double Insulation Symbol – a square within a square.

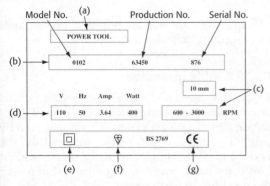

Fig 6.1 | Specification plate details

f   Standards mark – for example British
    Standards Kite mark and Number.
g   Products comply with the appropriate
    European Standards and Regulations.

6.2   Earthing, insulation & electrical safety

Here we must consider the different forms of
electrical supply, and how the operative is protected from the potential dangers associated
with them. For example:

- Tools using a mains electricity supply.
- Tools protected by Double Insulation.
- Tools using a 110volts supply
- Tools connected via a Residual Current
  Device (RCDs)
- Tools run off Portable Electricity Generators
- Safe use of electrically operated hand power
  tools.

6.2.1 Mains supply

Figure 6.2 shows how an electric drill or similar
machine is protected by an earth line (wire). The
principle is as follows. If the electrical appliance
has a low-resistance connection to earth, then a
return path is available for the current if a fault
occurs that causes the appliance's casing to
become live. This low-resistance path allows a
high current to flow which causes the fuse (providing it is correctly rated) to burn out (blow),
thus stopping the flow of current and rendering
the machine safe.

No matter how effective this system might
seem, there is always the possibility that it will
not work when it is most needed. For example,
the system will fail to work if the earth wire
has:

- not been connected to the plug socket, or
  has become defective en route or at source,
- become disconnected from the machine,
- become disconnected from the plug,
- been damaged within the flexible cable to
  the machine, or the extension lead.

If the earthing system fails to work, possibly for
one of the above reasons, it could result in the
operator's body being used as an electricity
escape route to earth – the results of which could
prove fatal.

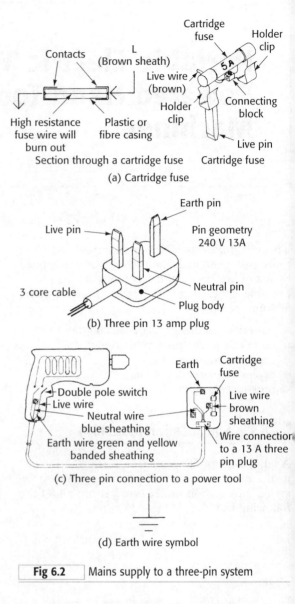

(a) Cartridge fuse

(b) Three pin 13 amp plug

(c) Three pin connection to a power tool

(d) Earth wire symbol

| **Fig 6.2** | Mains supply to a three-pin system |

Fortunately, nearly all portable power tools
produced are now *double-insulated*.

6.2.2 Double insulation

Figure 6.3 shows how a double barrier is formed
around all those components capable of conducting an electrical current. This is achieved by
using a strong non-conductive material for the
body and/or isolating any metal parts with a
non-conductive inner lining, thus eliminating
the need for an earth wire.

Portable power tools which are double insu-

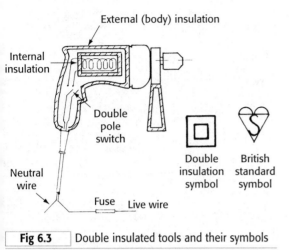

Fig 6.3   Double insulated tools and their symbols

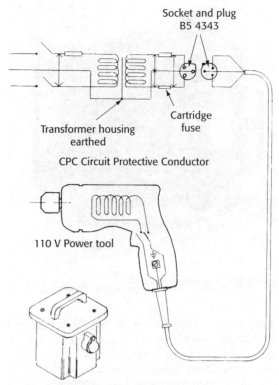

110 V Transformer (identification colour yellow)

Fig 6.4   110 V supply via a step down transform

lated bear the symbol of a square in a square and, before these tools can legally be used in industry, dispensing with their earth wire, they must comply with the Electricity at Work Regulations 1989 by being approved by the British Standards Institution and bearing BSI's 'Kitemark' BS 2745 and 2769 on the casing.

Double-insulated tools are undoubtedly safer than single-insulated tools (normal earthed systems) but, unless they are powered from a low voltage supply (110 V), there is still considerable danger from the current-carrying cable.

6.2.3 110v Power supply

This is the specified voltage for power tools used on a building site figure 6.4 shows how the supply voltage of 240 volts is reduced to 110 volts, with a centrally tapped earth, so that, if a breakdown in insulation does occur, the operative should only receive a shock from 55 volts.

6.2.4 Portable electrical generators

Were a mains supply is not available, one alternative is use a portable generator powered by either a petrol or diesel engine. A typical example is shown in figure 6.6. These are available with different power outputs. Figure 6.5 shows the electric plug/socket geometry.

Generators must be sited outdoors (in the open air) – exhaust fumes are toxic. It must be installed according to BS 7375 by a competent electrician and must be earthed.

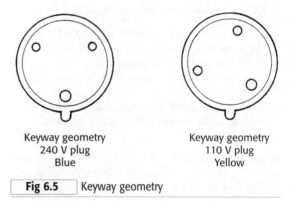

Keyway geometry   Keyway geometry
240 V plug   110 V plug
Blue   Yellow

Fig 6.5   Keyway geometry

6.2.5 Residual current device (RCDs) (fig 6.7)

These are devices which, in the event of an electrical fault occurring where current flows to earth. They work by detecting an imbalance in the flow of electricity between the live and neutral wires in the circuit such as would occur if

Fig 6.6 | Portable generator

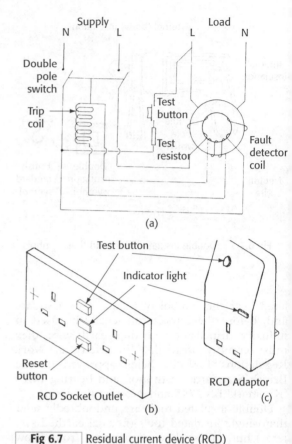

(a)

(b)

RCD Socket Outlet   RCD Adaptor (c)

Fig 6.7 | Residual current device (RCD)

there was a leakage to earth, in which case the double pole switch would disconnect the supply almost instantaneously (fig 6.7a). They do, however, depend on the proper functioning of their electrical and mechanical components, it is essential therefore, that they are tested at regular intervals (before each work period) – test buttons are provided to enable this to be carried out.

RCDs are available in several forms;

- One which is permanently combined with the plug and directly wired to the tool.
- Built into the mains consumer unit.
- A socket outlet in a ring main system (fig 6.7b).
- An adapter fitted between plug on the tool lead and the socket (should not be used – see below).

Note: RCDs adapters (fig 6.7c) should not be used as there is always the risk that they may accidentally be forgotten or left out of the system.

6.2.6 General safety

Before a portable power tool is used, the operator must be confident that all necessary steps have been taken to ensure both his/her safety and that of any persons within close proximity of the operation to be carried out.

The precautions listed below should always be followed:

1. Never use a portable power tool until a competent person has instructed you in its use.
2. Only use a portable power tool after authorised approval (the tool in question may have been withdrawn from use or some reason of safety).
3. The manufacturer's handbook of instruction for the tool in question should be read and understood before use.
4. Always wear sensibly fitting clothes – avoid loose cuffs, ties, and clothes which are torn etc.
5. Wear eye protection where there is a risk of propelled debris or fume, dust or mist which may impair or damage your eyesight.
6. The correct type of Dust masks should be worn where the operative's health may be at risk.
7. Ear protection should be worn by all operatives who are likely to be subjected to

a noise at 85dB(A) and above, for example if you have to shout to make yourself heard by someone 2 metres away. Examples of some portable powered hand tools where approved ear protection should be worn are: – Impact (percussion) drills, circular saws, planers, routers, and cartridge operated tools. Under the Noise at Work Regulations 1989 employers are required to reduce the risk of hearing damage.

8 Guards, where fitted, must always be used.
9 Never use blunt or damaged cutters.
10 Keep flexible cables away from the work-piece, cutters, and sharp edges and also from trailing on the floor.
11 Before changing bits, abrasive sheets or making any adjustments, always disconnect the tool from the electric supply (remove the plug from its socket).
12 If a tool is damaged or found to be defective, return it to the stores or to the person responsible for it. Ensure that it is correctly labelled regarding the extent of its damage or defect.
13 If injury should occur – no matter how minor – first aid must be applied immediately to avoid the risk of further complications. The incident should then be reported to the person responsible for safety.

Also see appendix 1.

6.3 Use of portable power tools

Portable electric powered hand tools are available to carry out the following functions;

a drilling,
b sawing,
c planing,
d rebating,
e grooving,
f forming moulds,
g screw driving,
h sanding.

Provided a suitable supply of electricity is within easy reach, and the amount of work warrants their use, these tools can be used to speed up hand operations and where the use of permanent machinery would be impracticable.

Some portable power tools are produced specifically for the 'do-it-yourself' market and should not be confused with 'industrial tools' – although they may appear the same, differences occur in both cost and the inability of do-it-yourself tools to sustain constant industrial use. For example:

Type	Category	Use
Light duty	Do-it-yourself	Occasional
Medium duty	General-purpose: tradesman	Moderate to intermittent
Heavy duty	Industrial	Continuous

6.4 Electric drills (rotary)

Choice of an electric drill will largely depend on:

a type of work,
b volume of work,
c size of hole,
d type of material (wood, metal, soft brick or blockwork etc).

Figure 6.8 (a) shows a typical palmgrip drill. Figure 6.8 (b) shows a heavy duty four-speed 'D' back handled rotary drill.

Ideally, an electric drill should be adjusted to rotate at a speed to suit both the material (workpiece) and the hole size (drill bit). Provided the drill is powerful enough and the correct drill bit (see table 6.1) is used, it is possible to bore holes in most materials (including soft masonry).

6.4.1 Drill Cutting Speeds

A drill bit should rotate at its most effective cutting speed, otherwise it could overheat, quickly become dull, or even break. Cutting speed is often misquoted as revolutions per minute (rev/min), which only denotes the number of times the chuck revolves every minute (the drill speed). To determine a drill bit's cutting speed (edge speed of bit), the distance it travels every revolution must be known.

As can be seen from figure 6.9, the distance covered by one revolution will vary according to the diameter of the bit.

Therefore, to find the cutting speed of a drill bit – which is usually quoted as metres per second (m/s) – we use the following formula:

Cutting speed (m/s) = $\pi \times$ Diameter (m) $\times$ Drill speed (rev/s)

where π ('pi') = 22/7 or 3.142.

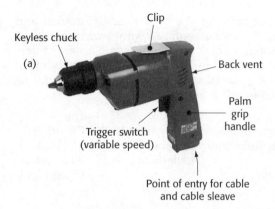

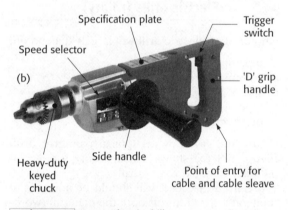

Fig 6.8 | Rotary electric drills

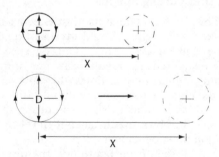

D = diameter of drill bit
X = Distance covered by the cutting periphery
(outer edge of the drill in **one revolution**)

NOTE: If the drill speeds in RPM are constant a point on
the outer edge of the larger diameter drill will travel
further in the same time as shown at X

Fig 6.9 | Drill bit cutting speed

For example:

For a drill bit of 6 mm diameter in a drill with a working speed of 1500 rev/min the distance around the outer edge (circumference) of the drill would be: $\pi \times$ Dia. or (πD)

1 First let us change the diameter from millimetres to metres by dividing by 1000

so: $\dfrac{6}{1000} = 0.006$m

Resulting in; $3.142 \times 0.006 = 0.018852$m

2 Now convert RPM to number of revolutions (rev) per second(s):
Divide the spindle speed (RPM) by 60 to reduce to seconds:

$$\dfrac{\text{Spindle speed (RPM)}}{60}$$

$$\dfrac{1500}{60} = 25\text{rev/s}$$

3 The cutting speed at metres per second (m/s) can now be determined by using :

0.018852m $\times 25$rev/s $= 0.47$m/s

or $\dfrac{3.142 \times 6}{1000} \times \dfrac{1500}{60} = 0.47$m/s

When drilling holes in metals a guide to the recommended ranges of cutting speeds is shown in table 6.1.

Drilling holes in wood usually requires about twice the cutting speed for metal, but, because of the many hard and abrasive wood-based materials now in common use, consideration should be given to the material's composition before choosing a bit or its speed.

Table 6.1 | Guide to recommended cutting speeds

Material	Cutting speed (m/s)
Aluminium	1.00 to 1.25
Mild steel	0.40 to 0.50
Cast iron	0.20 to 0.40
Stainless steel	0.15 to 0.20

Drill manufacturers usually recommend the appropriate rev/min to suit both the drill bit size and the material. *It is a good idea to take a copy of these and attach it to (ideally inside) the drill carrying case.*

Note: *As a general guide, the harder the material or larger the hole the lower the rev/min.*

pilot holes (fig 12.7) should be pre-bored before driving any wood-screw.

Figure 6.16 shows a screwdriver with automatic feed. Screws are fed to the driver via a strip of plastic into which they are temporary attached.

Note: because cordless drills also double up as power screwdrivers they are more commonly used – particularly on site work (see section 7).

6.9   Sanders

There are many different types and styles of sanders associated with bench work. We shall be looking at three of these, namely:

- belt sanders
- random orbit disc sanders
- orbital sanders

6.9.1 Belt sanders

Designed and constructed to tackle heavy sanding problems with minimum effort. Figure 6.17(a) names the main parts of this machine. The endless abrasive belt is driven by a motor-driven rear (heel) roller over a front (toe) belt-tensioning roller and then over a steel-faced cork or rubber pressure plate on its base – this is the part that makes contact with the workpiece. Dust is discharged via a suction-induced exhaust into the dust bag.

a *Method of use* – It should be permitted to reach full speed before being gently lowered on to the workpiece – allowing its 'heel' to make contact slightly before its 'toe', to avoid any kickback. When contact is made, there will be a tendency for the sander to move forward and force the workpiece backwards, due to the gripping action of the sanding belt (see fig. 6.18), so both the sander and workpiece must be held firmly at all times. The surface finish produced by the sander will depend on what grade of abrasive belt has been used. Figure 16.17(b) shows a belt sander being used with a sanding frame attachment. (See section 5.10.2; table 5.6 and fig. 5.86).

Note: the dust bag must remain attached to the sander while the motor is in motion, otherwise dust and particles of grit will be discharged directly from the exhaust tube at an alarming rate and could result in serious injury.

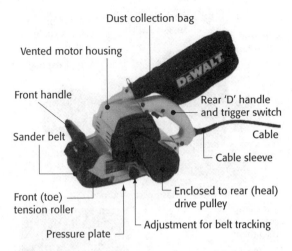

(a) Main components

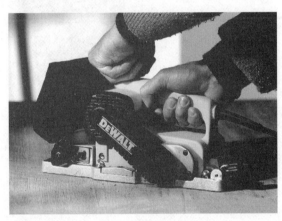

(b) Belt sander in use with a sanding frame.

| **Fig 6.17** | Portable belt sanders |

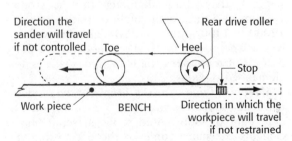

| **Fig 6.18** | Securing the workpiece being sanded |

6.9.2 Orbital sanders and finishing sanders

Figure 6.19 shows the main parts of a heavy duty machine. The abrasive sheet is attached to the

Front grip
(90° rotation)

Back handle with
trigger switch

Point of entry for
cable & cable
sleeve

Dust bag

Rear clamp lever–
abrasive paper
attachment

Base plate and pad

Rear clamp lever–
abrasive paper
attachment

Orbital sander in use with
attached dust extraction
hose

Fig 6.19   Orbital ½ sheet heavy duty sander

On/Off
switch

Point of entry for
cable & cable
sleeve

Specification
plate

Dust bag

Punch base plate

Levers for abrasive
paper grip

Fig 6.20   Orbital 'Palm' ¼ sheet finishing sander

6.9.3 **Random orbital disc sander** (fig 6.21)

Single and double handed machines are available, capable of producing excellent finished results by working with an eccentric (off-centre) action whilst rotating. The base is circular with a flexible sanding pad, which, like the abrasive disc which is attached to it, is perforated allowing machined dust to be drawn into a collection bag or extraction system. Self-adhesive sanding discs with peel-off backing, or 'Velcro' type backed sanding discs are usually available.

Method of Use – application methods will be similar to all finishing sanders.

6.9.4 **Dust extraction**

As with all mechanical-sanding operations, dust can be a serious health hazard, and, even though all belt sanders and orbital sanders are fitted with a means of dust extraction (extraction methods do differ between models), an approved mouth and nose mask must be worn by all operatives and those in close proximity, whenever sanding operations are being carried out.

6.10   Circular saws

Figure 6.22 shows a typical portable circular saw. Its main components have been identified

base plate either via clips, levers, or are 'Velcro' (depending on the design and dust extraction system), the orbital diameter can be from 1.5 mm to 5 mm, with rotation from between 12000 to 1400 rev/min (RPM). These may also be known as ½, ⅓ or ¼, sheet sanders (see fig. 6.20). **Palm sanders** (fig 6.20) are smaller single handed versions.

Method of use – the self-weight of the sander is usually sufficient pressure – anything other than light pressure could result in scratching, clogging the paper, or even the body orbiting while the pad appears to remain stationary. To obtain the best results, the appropriate grade of abrasive paper must be used to suit the job. This often means starting with a coarse grade of paper, then reducing the grade until the desired result is achieved (see section 5.10.2, tables 5.6 & 5.7 and fig. 5.86).

On/Off switch

Palm grip (see below)

Cable sleeve

Sanding pad

Detachable dust bag

Typical application

Fig 6.21 Random orbital 'Palm' grip sander

Carry and operational handle

Aperture for chip and sawdust extraction

Motor and motor housing

Front handle

Top (crown) guard– arrow showing blade rotation

Riving knife

Base (sole) plate

Retractable self-closing blade guard

Adjustable fence

Anti-friction rollers on retractable blade guard

Scale and adjustment for bevelled cutting

Back handle and trigger switch

Point of entry for cable and cable sleeve

Front view with blade set at an angle

Motor

Base (sole) plate

Blade height adjustment bracket and locking device.

Back view of saw with blade lifted above base plate

Fig 6.22 Portable circular saw

by name and/or function. Saws like this are capable of cutting a variety of materials, such as

a timber:
   - softwood,
   - hardwood.
b manufactured boards:
   - veneer plywood,
   - core plywoods (blockboard and laminboard),
   - particle board (chipboard etc),
   - OSB (Orientated Strand board),
   - fibre boards (hardboard and insulation board),
   - MDF (Medium Density Fibreboard) – special attention should be given to dust extraction (section 6.9.4) and prevention of dust inhalation.
c plastic laminates

Provided the correct blade for the material being cut is used as recommended by the saw's manufacturer, a very satisfactory cut can be made. Most blades are now tipped with Tungsten

Carbide (known as TCT blades), these can be broadly broken down into four groups:

- Ripping
- General-purpose
- Cross-cutting
- Fine cross-cutting

The cutting edge of these blades last much longer than traditional steel plate blades.

Although there are many variation on blade make and arrangement of the teeth, generally speaking the greater the number of teeth around the circumference of the blade (tooth pitch – for more information see section 9.5.1) the finer the cut. For example, if we consider two blades of the same diameter the one used for ripping would have less teeth than the one used for crosscutting. Whereas blades with a mid-range number of teeth may be classed as a general-purpose blades.

The efficiency of the blade, and to some extent the safety of the cutting operation, will depend on the sharpness of the blade. Blades must therefore be kept sharp. TCT blades require specialist sharpening equipment and therefore have to be returned either to the manufacturer or their approved agent for re-sharpening.

Blade replacement must be carried out according to the toolmaker's instruction, as this process may differ slightly between makes.

Blades cut on the upward stroke and rotate in the direction of the arrow (fig 6.28) – usually shown on the top guard.

6.10.1 Methods of use

Portable circular saws are capable of carrying out the following operations:

i   *Ripping* – Figure 6.23 shows how the same ripping guide is used and how, by using a temporary fence, wide boards can be sawn.

ii *Rebating* – *Two* cuts are required, as shown in figure 6.24.

iii *Cutting bevels with the grain* (fig. 6.25) – any angle between 90° and 45°;

iv *Cross-cutting* – Accurate cuts can be made by using a jig like the one shown in figure 6.26.

v   *Cutting bevels across the grain* – any angle between 90° and 45° by modifying the jig shown in figure 6.26 or by using a 'Pull Over Mitre Saw' (fig 6.29)

vi *Cutting compound bevels* – Use a temporary fence, or a jig like that shown in figure 6.26 and saw angle adjustment, preferably a 'Pull Over Mitre Saw' (fig 6.29) should be used. The cut resembles face and edge cuts to a roof jack rafter (see Book 2 section 8).

vii *Cutting mitres* – Use a temporary fence or a jig and angle adjustment – preferably a 'Pull Over Mitre Saw' (fig 6.29) should be used.

viii *Cutting plough grooves* – Make a series of cuts with the grain, using the aids mentioned.

ix *Trenching (housing)* – use a jig as shown in figure 2.26 or 6.27.

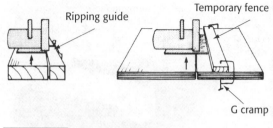

Fig 6.23   Ripping operation

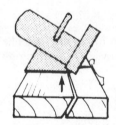

NOTE: Before the rebating cut is completed provision must be made to prevent the waste portion being shot forward due to the forward motion of the saw blade

Fig 6.24   Grooving and rebating

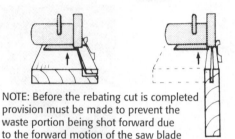

Fig 6.25   Bevel cutting with the grain

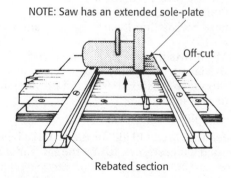

NOTE: Saw has an extended sole-plate

Fig 6.26   Cross cutting with a jig

cut gives a sufficiently fine finish that mitres and abutments would require no further treatment (planing).

As with most other makes and models there is provision for connecting a dust extraction unit to the machine.

6.11.2 Safe operation

The same rules of safety apply to these machines as to other woodworking machines as featured in chapter 9. See also the general safety rules in section 6.2.6 and appendix 1.

6.12 Combination saw bench and Mitre saws

Figure 6.30 illustrates an easily transportable 'flip-over' saw unit with detachable legs, which as shown can covert easily from a small saw bench into a mitre saw. *It is vitally important that at each change over (from a saw bench to a mitre saw and visa versa) the special locking devices are fully set in place and secured.*

Machines of this type are generally capable of ripping down small-sectioned timber, crosscutting, cutting mitres and making bevel cuts. Several attachments are generally available, for example, to extend the saw table etc.

Note. There are other makes that carry out similar functions.

6.12.1 Use

The saw bench aspect of this machine can rip down short lengths of timber up to 70 mm in thickness. The mitre saw set-up could also straight crosscut timber up to 210 mm wide. Maximum dimensions of material for other operations will depend on the angle/bevel being cut.

6.12.2 Safe operation

The same rules of safety apply to these machines as others See also the general safety rules in section 6.2.6 and appendix 1.

Note: many portable circular saw benches in their own right are readily available. Their use must comply with current legislation for woodworking machines as featured in chapter 9.

6.13 Reciprocating saws

Saws that fall into this category cut by the reciprocating (backward and forward or upward and downward movement) action of their blade.

For the purpose of this chapter I have divided these into three categories:

1 Forward pointing *wide* bladed reciprocating saws
2 Forward pointing *narrow* bladed reciprocating saws
3 Jig saws

6.13.1 Forward pointing wide bladed reciprocating saws (fig 6.31)

These saws are also known as:

● Powered hand saw, or
● Alligator saw.

They work by using two counter acting reciprocating replaceable blades. The type of blade will depend on the intended use of the saw. For example, cutting timber, manufactured boards or building blocks.

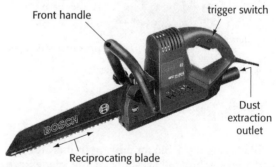

Front handle

trigger switch

Dust extraction outlet

Reciprocating blade

| **Fig 6.31** | Forward pointing wide bladed reciprocating saw

6.13.2 Forward pointing narrow bladed reciprocating saws (fig 6.32)

These saws are also known as:

● All-purpose saw,
● Reciprocating saw,
● Sabre saw, or
● Shark saw.

Some of these saws have provision for a reciprocating action similar to jig saws.

Specification plate

Reciprocating and orbiting blade

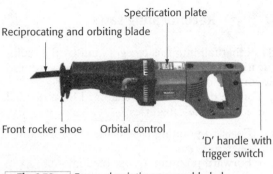

Front rocker shoe   Orbital control

'D' handle with trigger switch

Fig 6.32 Forward pointing narrow bladed reciprocating saw

Handle with trigger switch

Switch lock (on) –access from both sides

Electronic speed control dial

Cable sleeve

Clear view blade guard –chip deflector

Motor

Tilt lever lock for base plate

Base plate

Dust extraction outlet

Reciprocating blade

Orbital action selector

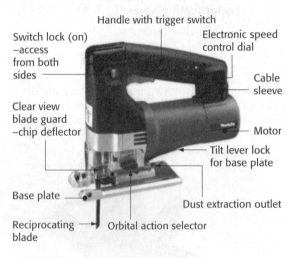

Fig 6.33 Jig saw

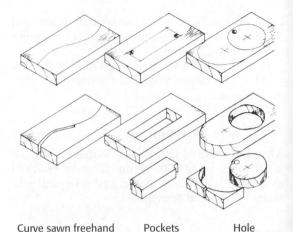

Curve sawn freehand   Pockets   Hole

Fig 6.34 Cutting process

Fig 6.35 Using a jig saw with a circle cutting attachment

Blades are available for cutting timber, ferrous and non-ferrous metals. One of the blades can be used for cutting through wood which may contain embedded nails, making the saw ideal for maintenance, renovation, and refurbishment work.

6.13.3 **Jig saws** (fig 6.33)

Although capable of cutting a straight line – usually with the aid of either a parallel guide attachment or a temporary fence – the main function of jigsaws is cutting slots, curves, and circles, as shown in Figure 6.34. Using an attachment like the one shown in figure 6.35 the saw can cut circles. A rip fence is used for making parallel cuts.

Most models incorporate in their design a mechanism that directs the blade into an orbital path (pendulum action). This additional facility can be used to improve the cutting action of the blade when sawing wood, by allowing the blade to clear sawdust from the kerf on its non-cutting stroke, thereby reducing unnecessary friction.

Figure 6.36 gives an impression of how the orbital action operates. The degree of orbiting will depend on the material being cut – as indicated. Straight reciprocating action is used for metals, and the amount of orbiting is increased to suit the degree of softness of the material being cut – the softer the wood, the greater the

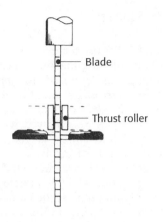

Front elevation

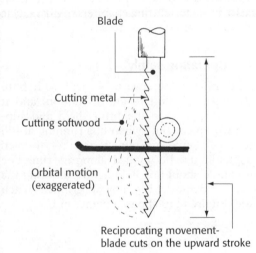

Side elevation

| **Fig 6.36** | Orbiting (pendulum action of saw blade) |

| **Table 6.3** | Orbital action for different cuts |

Dial position	Cutting action	Material application
0	Vertical action	For cutting mild steel, stainless steel, plastics Fine cuts in wood & plywood
I	Small orbit	Mild steel, aluminium and hardwood
II	Medium orbit	Softwood, plywood – fast cutting aluminium
III	Large orbit	Fast cutting softwood

orbit – table 6.3 can be used as a guide to cutting different materials (specifications may vary between makes).

Provided the correct blade is used, a wide range of materials can be cut. Only those blades made or recommended by the manufacturer of the tool should be used, and a specification for blade selection is issued with each tool. Table 6.4 gives some idea of the kind of performance to be expected from a blade, but it must be stressed that this is only a guide (specifications may vary between makes). Figure 6.37 shows two different blade fittings.

| **Table 6.4** | General guide to jigsaw blade suitability |

Material to be cut	Blade details*		
	Max. thickness of cut*	Teeth per 25 mm	Tooth pitch |mm|
Softwood and hardwood	30 mm	10	2.5 mm
Softwood and hardwood	60 mm	6	4 mm
PVC and acrylics	13 mm	8	3 mm
Aluminium	16 mm	8	3 mm
†Mild steel and aluminium	1.5 to 4 mm	20	1.2 mm
Plywood and hardwood	16 mm	20	1.2 mm
†Mild steel	3 to 6 mm	14	1.8 mm
Alumium	10 mm	14	1.8 mm
Plywood and hardwood	32 mm	12	2 mm
†Stainless steel	1 mm	24	1 mm

*Details may vary between makes to tool.
†Special blades.

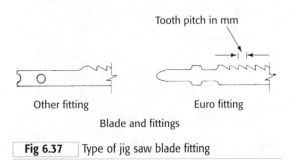

Blade and fittings

| **Fig 6.37** | Type of jig saw blade fitting |

6.13.4 Cutting operation & Safe use

After ensuring that the workpiece is fully supported and held firm, select the correct blade, speed (high speeds for wood, low for metals and plastics), and motion (straight or orbiting (pendulum) action – table 6.3). Depending on the type of work being carried out, cutting can start from one edge or after inserting the blade through a pre-bored hole, as shown in figure 6.34. It is not advisable to make a 'plunge cut' (fig. 6.38), as is sometimes practised in soft materials, because of the possibility of machine 'kickback' when the blade first makes contact with the surface. The method may also result in the blade being damaged or broken.

Apart from the general safety precautions, which must be followed at all times, operatives must be constantly aware of the unguarded blade – particularly that portion which protrudes below the saw cut.

See also the general safety rules in section 6.2.6 and appendix 1.

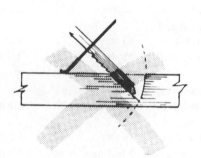

Plunge cut–*not* recommended

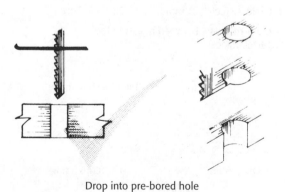

Drop into pre-bored hole

Fig 6.38 | Starting a cut midway

6.14   Planers

Planers (fig 6.39) use a rotary cutter block, similar to a large planing machine, thereby producing chippings as opposed to shavings. This makes them especially suitable for removing waste wood quickly – cuts of up to 3 mm deep with one pass can be achieved with the larger models. If a fine finish is required, simply decrease the depth of cut by turning the front hand pre-select dial and move the planer over the surface at a rate which is sufficient to prevent a build-up of chippings at the chipping outlet.

Apart from surfacing and rebating both softwood and hardwood, some models are capable of safely cutting bevels and chamfers (one or two 'V' grooves are usually formed into the front sole as a guide for cutting chamfers). Most tackle awkward and end grain with ease.

6.14.1 Operating safely

The machine must always be held with both hands when the cutters are in motion, and it should be applied to the firmly held and supported workpiece only when it is running at full revolutions. Once in contact, the 'toe-to-heel' principle should be applied along its run. The front handle should be held lightly, allowing the machine to be pushed by a firmly gripped switch handle. Figure 6.40 shows a planer in use.

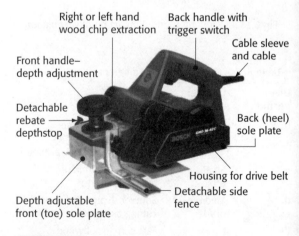

Right or left hand wood chip extraction

Back handle with trigger switch

Cable sleeve and cable

Front handle–depth adjustment

Detachable rebate depthstop

Back (heel) sole plate

Housing for drive belt

Detachable side fence

Depth adjustable front (toe) sole plate

Fig 6.39 | Portable powered planer

After switching the machine off, always allow the cutters to stop revolving before putting the machine down in the rest position. Protect the cutters at all times by making sure they are kept clear of any surface or obstacle (*some models incorporate in their design a drop-down sole rest, others may have a drop-down retractable cutter guard*). It is very important that the blades are kept sharp, otherwise the motor will suffer from overloading and the finish obtained will be poor.

Edging timber

Cutting a chamfer by using the middle 'V' groove in the sole as a guide

These machines require a safe means of collecting spent chippings – on the upper illustration this has been omitted to illustrate the hazard associated with chipping discharge

Fig 6.40   Portable powered planer in use

__Note:__ Cutters and moving parts must never be touched for any reason whatsoever until the machine has been unplugged and completely disconnected from its power supply.

See also the general safety rules in section 6.2.6 and appendix 1.

6.15   Routers

A router consists of a cutter rotating at between 18 000 and 24 000 rev/min, being driven by a vertically mounted motor set in a flat-bottomed framework.

Figure 6.41 shows a 'plunge' router, which, unlike older fixed-frame models, allows the cutter to be plunged vertically into the workpiece and retracted on completion of its task. This means

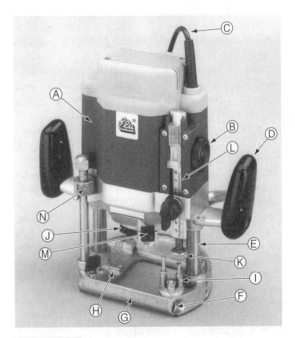

Fig 6.41   Plunge router
A – motor and housing; B – depth stop; C – cable and cable sleeve; D – side handles; E – plunge guide bars; F – location holes for fence rods; G – router base; H – template guide housing; I – swivel stops (three depth settings); J – collet chuck (bit holder); K – stop buffer (controls depth of plunge); L – depth scale; M – spindle lock (used when changing cutters); N – adjustable return stop

that the cutter need never be unduly exposed before or after the operation, thereby reducing the risk of damaging the cutter and of accidents.

Provided a suitable cutter is used, and in some cases special attachments, the following operations should be possible:

- cutting grooves – straight, curved, or circular
- cutting rebates into straight, curved, or circular shapes
- cutting slots and recesses
- cutting beads, and moulding onto straight, curved or circular shapes
- cutting dovetail joints,
- cutting dovetailed slots and grooves
- edge trimming plastics and wood veneers.

6.15.1 Router bits and cutters

For general wood-cutting these are made of high-speed steel, but a longer cutting life will be obtained by using bits and cutters which are tungsten carbide-tipped. Tungsten carbide (one of the hardest materials made by man) is bonded to an alloy-steel body to form a cutting edge capable of handling abrasive materials like plywood (hard glue lines), particleboards, and plastics. A few bits in the small range are available in solid carbide.

Table 6.5 in conjunction with figures 6.42(a) to (n) will give you some idea of the kind of work of which these bits and cutters are capable. Edge-forming bits usually include in their design a 'pilot' which allows the cutter to follow a predetermined path – either straight or curved. Pilots can be in the form of either a 'guide pin' or a 'guide roller'. With the guide-pin type of bit, the cut is restrained by the pin rubbing against the side of the wood, but movement is much easier with a roller, because of its ball bearing and lower friction.

6.15.2 Standard accessories

Usually include the following:

a a guide holder with
  - a straight fence for making parallel cuts (fig. 6.43)

Table 6.5 Guide to router bits and cutters

Bit/cutter type	Operation/ cutting	Mould/work profile	Figure 2.17
Grooving bits	Ploughing (housing/ trenching)		(a)
	Round core-box-bit groove		(b)
	Dovetail groove		(c)
	'V' groove		(d)
Edge-forming bits	Rounding over		(e)
	Beading (ovalo)		(f)
	Coving		(g)
	Chamfering		(h)
	Ogee		(i)
	Rebating		(j)
Trimming	Square edge (flush)		(k)
	Chamfer (bevel)		(l)
Edge cutter	Sawing		(m)
	Slotting		(n)

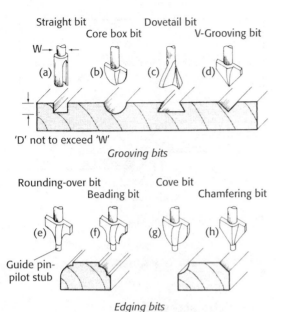

Straight bit

Core box bit

Dovetail bit

V-Grooving bit

W→

(a)   (b)   (c)   (d)

'D' not to exceed 'W'

Grooving bits

Rounding-over bit

Beading bit

Cove bit

Chamfering bit

(e)   (f)   (g)   (h)

Guide pin-
pilot stub

Edging bits

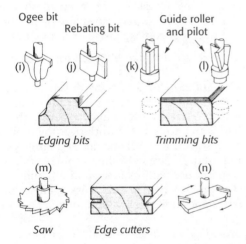

Ogee bit

Rebating bit

Guide roller
and pilot

(i)   (j)   (k)   (l)

Edging bits   *Trimming bits*

(m)   (n)

Saw   *Edge cutters*

Fig 6.42   Router bits and cutters

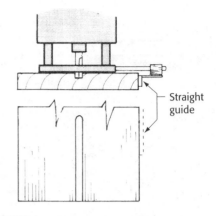

Straight
guide

Fig 6.43   Straight and Roller guides

b a template guide for reproducing identical cuts or shapes – examples are shown in figure 6.44;

c router bits (fig. 6.42);

d tools – spanners etc.

6.15.3 Operating the router

Having ensured that the workpiece is held securely, the router can now be picked up (holding it firmly with both hands) and positioned on or against the workpiece – depending on the type of machine (plunge or fixed-frame) and the work being done. Make sure that the bit/cutter is free to rotate, then start the motor. Allow it to reach maximum revolutions before making the first trial cut (usually in the waste wood or on a separate test-piece).

Deep cuts should be made by a series of shallow cuts – never cut deeper than the bit is wide (fig. 6.42 (a)), otherwise too much strain will be put on both the bit and the motor.

The cut is made from *'left'* to *'right'*, to allow

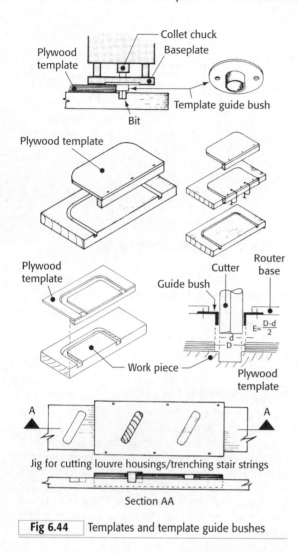

Jig for cutting louvre housings/trenching stair strings

Section AA

Fig 6.44 | Templates and template guide bushes

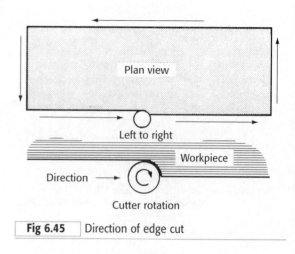

Plan view

Left to right

Workpiece

Direction

Cutter rotation

Fig 6.45 | Direction of edge cut

the bit/cutter to cut the material against its turning circle (see fig. 6.45). Move the router just quickly enough to make a continuous cut – never overload the motor (listen to its drone) nor put unnecessary strain on the bit by pushing too hard (fast). Moving the router along too gently (slowly) will allow the cutting edge and the wood to generate heat (by friction) which could result in the bit/cutter 'blueing' (overheating) and wood being burnt. A dull (blunt) bit/cutter will have the same effect.

At the end of each operation, switch off the motor. The bit/cutter should be free of the work and allowed to stop revolving before the machine is left.

6.15.4 Extra accessories

The following accessories are particularly useful for small firms, where specialised machines would be either too expensive or impracticable.

a *Trammel point and arm* – used for cutting circles.
b *Dovetail kit* – for producing dovetail joints.
c *Bench stand* – converts a router into a small yet practical spindle moulder, leaving the hands free to feed the work over the table. A push stick (fig. 9.10) and cutter guards will be required. Figure 6.46 shows router table and stand in use with its fence and top and side shaw guards (which hold work against the fence and table) in position forming a tunnel guard.
d *Trimming attachment* – used to cut veneered edging flush and/or bevelled.

6.15.5 Edge trimmers

Machines specially designed to trim overhanging edges of veneer. Figure 6.47 shows an edge trimmer being used to trim the edge of a laminate veneer.

Safety Notes: *Throughout all routing operations the following precautions should be taken:*

● Because there is always a risk of flying particles entering the eyes, approved goggles, face visor or safety glasses must be worn by the operative.

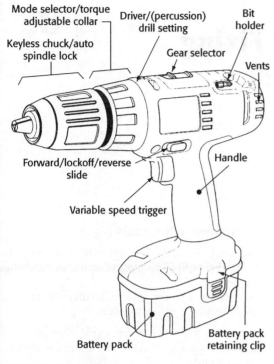

Mode selector/torque adjustable collar

Driver/(percussion) drill setting

Bit holder

Keyless chuck/auto spindle lock

Gear selector

Vents

Forward/lockoff/reverse slide

Handle

Variable speed trigger

Battery pack

Battery pack retaining clip

Fig 7.2 The main parts of a cordless Driver/(percussion) Drill (DeWALT)

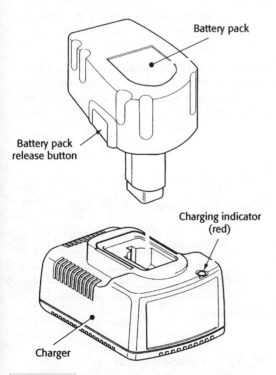

Battery pack

Battery pack release button

Charging indicator (red)

Charger

Fig 7.3 Battery pack and charger (DeWALT)

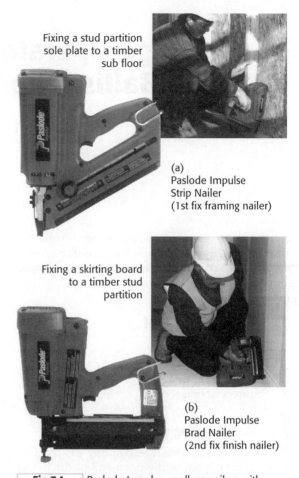

Fixing a stud partition sole plate to a timber sub floor

(a) Paslode Impulse Strip Nailer (1st fix framing nailer)

Fixing a skirting board to a timber stud partition

(b) Paslode Impulse Brad Nailer (2nd fix finish nailer)

Fig 7.4 Paslode Impulse cordless nailers with examples of application

skirting boards, architraves, dado rails, paneling, and a various trims.

It is capable of driving nails with lengths from 19 to 64 mm with a magazine capacity of up to 150 nails, depending on their size. It can (depending on fixing type) drive a maximum of 2 nails per second and 500 per hour.

7.2.3 Safety

When you are using these tools, eye protection must be worn and training in the safe handling and correct use of the tool must be given this is provided via the manufacturer. The tool has usual built in safety devices, but it must be used according to the instructions provided in the manufacturer's literature.

Note: *The relevant part of section 7.1.2 will also apply.*

Cartridge Operated Fixing Tools (Ballistic Tools)

8

These tools use compressed gas from an exploded charge, which has been contained in a cartridge, to propel and drive a fixing device into a base material – possibly via a workpiece.

8.1 Types of tool

Basically there are two types:

i   direct-acting tools and
ii indirect-acting tools.

8.1.1 Direct-acting tools (fig. 8.1)

These use the contained expanding gas to drive a fixing device out of the open end of the barrel at a high velocity. They can therefore be classed as high-velocity tools.

8.1.2 Indirect-acting tools (fig. 8.2)

These have a captive piston which intervenes between the explosive charge (cartridge) and the fixing device.

Depending on the manufacturer there are three methods of operation (fig. 8.3):

a the piston is forced by the expanding gas down the barrel to strike the fixing device at the exit end;
b the piston and fixing device are in contact and are both forced down the barrel together;
c the piston and fixing device are in contact at the exit end and driven together while in contact with the work face.

The style of the tool generally resembles a pistol, as shown in figure 8.4 with a firing mechanism that consists of a spring-loaded firing pin operated by a trigger.

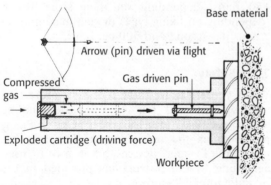

Fig 8.1 Direct-acting tools (high velocity types)

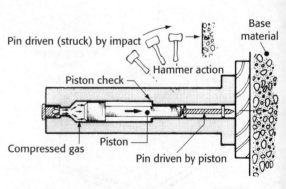

Fig 8.2 Indirect-acting tools

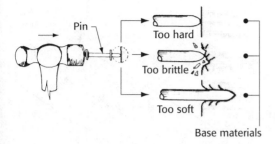

Fig 8.11 Testing for a suitable base material

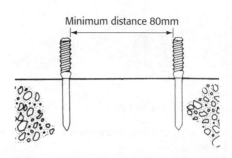

Fig 8.13 Minimum distances between fixings into concrete

8.5 Fixing to concrete

Nails and studs driven into concrete generate a very high pressure temperature up to 900°C around the point of the pin. As a result the concrete and steel sinter (fuse together with the fixing pin) which produces a holding powers of up to 15 kn. However, for this reaction to work satisfactorily, the following points must be observed:

- Establish the suitability of the base material, as previously discussed.
- Ensure a suitable depth of penetration – by using the correct cartridge strength (fig. 8.12).
- Always leave a minimum gap of 80 mm between fixings (Fig. 8.13).
- Ensure that the base material is thick enough to give a satisfactory penetration -as shown in figure 8.14.
- Never fix nearer than 80 mm to the edge of a crater caused by surface lifting ('spall' – fig. 8.15).
- Never fix closer than 80 mm from a free

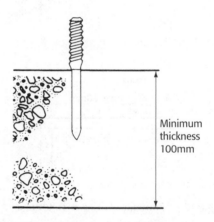

Fig 8.14 Minimum thickness of concrete

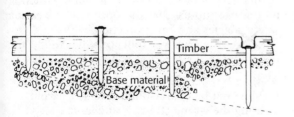

Fig 8.12 Examples of how cartridge strength may affect penetration of the base material

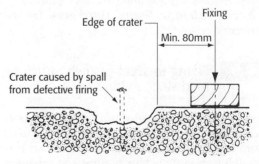

Fig 8.15 Proximity of fixings to a crater edge

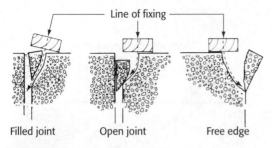

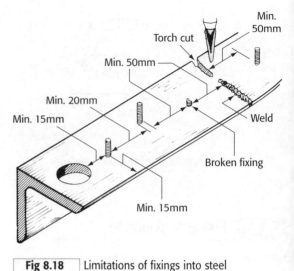

Fig 8.16   Possible line of deflection to a joint or free edge of concrete

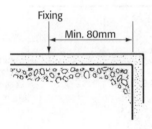

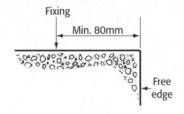

Fig 8.17   Proximity of fixing to concrete joint or free edge

Fig 8.18   Limitations of fixings into steel

edge or joint (fig. 8.16) – figure 8.17 shows minimum requirements.

● Never fix into mortar joints.

Note: Always use a splinter (safety) guard.

Never attempt fixing into pre-stressed concrete.

8.6   **Fixing to steel** (usually structural mild steel sections)

Fixings driven into steel should, where possible, pierce the steel as shown in figure 8.7 for maximum holding power. Fixing situations without through penetration will require special attention, as the reactive compressive forces around

the pin may tend to push the fixing back out of the steel.

When fixing into steel, the following points must be observed:

i   The tool manufacturer should be consulted if there is any doubt as to the suitability of the steel as a base material.

ii   Never drive a fixing closer than 15 mm to a free edge or hole (fig. 8.18).

iii The distance between fixings should be not less than 20 mm (fig. 8.18).

iv   Never drive a fixing closer than 50 mm to either a broken-off pin, a weld, or, where the steel has been cut with a blowtorch (fig. 8.18).

8.7   **Safe operation**

8.7.1   **The operative**

Figure 8.19 shows a low-velocity indirect-acting tool in use (notice that the operative is wearing safety goggles and ear protection).

Before a cartridge tool can be used, the operative:

● Must be over the age of 18 years.
● Must not be colour-blind – so as to avoid using the wrong strength of charge (cartridges are colour coded according to their explosive power).
● Must have been trained in its use by the

| Fig 8.19 | 'Hilti' low velocity tool in use |

manufacturer's training instructor and have obtained a certificate of competence.

Note: Because there are differences between makes and models of ballistic tool, each often requires a separate certificate of competence.

- Must be competent in its use and aware of any potential hazards which might arise, be capable of acting accordingly – for example, a 'misfire' (see section 8.7.2) must be dealt with as stated in the manufacturer's instructions for use.
- Must be suitably dressed, wearing the correct safety equipment, helmet, goggles, and ear protection.
- Must be aware of the dangers of working near flammable vapour, or in an explosive atmosphere.
- Must be aware of the possibility of a recoil (kickback) from the tool – never work from ladders, as both hands are required to operate the tool correctly. Good balance should be sustained at all times during the operations – loss of balance could prove fatal.
- Must never load a cartridge into the tool away from where it is to be fired, or carry the tool about when loaded with a cartridge.
- Must be aware of the dangers to bystanders, onlookers, or passers by – for example, the possibility of a ricochet, which could put people in close proximity at risk of injury.
- Must, where there is any likelihood of a fixing piercing a base material, make

provision for the area behind the wall etc. (usually blind to the operator) to be screened off so as to totally eliminate any risk of injury from either a fixing or a projected particle of the base material.

Note: The manufacturer's instructions must be observed at all times. It is important, therefore, that they are always available to the operative (kept with every tool) and that they are understood – ambiguities should be clarified by either the manufacturer or an appointed agent.

8.7.2 Misfire

A misfire may be as a result of a faulty firing mechanism or cartridge. If a misfire does occur, take the following action:

- keeping the tool held up against the work face, reactivate the trigger mechanism and fire the tool again
- if after taking the above action the tool still fails to fire, wait 30 seconds then remove the tool from the work face; keeping the muzzle pointing down and away from you remove the cartridge by following the manufacturers instructions
- correctly replace the tool into its metal carrying case clearly labelling the fault, and return to the maker or an appointed agent for repair.

8.7.3 Servicing

To keep the tool in good working condition, it will require regular cleaning and servicing in accordance with the manufacturer's instructions.

Any tool found to be defective must be removed from service and returned to the person responsible for its safe keeping and maintenance, clearly stating the nature of the fault, so that it can be clearly labelled accordingly before being returned to its maker or an appointed agent for repair.

8.7.4 Security

Cartridge-operated tools and their accessories should always be kept with the maker's case, and under the control of authorised operatives. After use, they should be kept unloaded in a locked container.

Checks on case contents should be made on

collection from the store and on their return, by the storekeeper.

Spare cartridges must be kept in their correctly coded containers – kept in a compartment in the metal container – never loose.

8.7.5 References and recommended further reading

- British Standard BS 4078: 1966/1987, 'Cartridge-operated fixing tools'.

- Health and Safety Executive Guidance Note PM 14, 'Safety in the use of cartridge – operated fixing tools'.
- Powder Actuated Systems Association, *Guide to basic training.*
- Manufacturers handbooks giving instruction on use and application.

Basic Static Woodworking Machines

9

Edited and updated by Eric Cannell

The aim of this chapter is to help the student to become aware of the more common types of woodworking machinery that the carpenter and joiner may encounter, and to be able to recognise these machines by sight and to understand their basic function.

Undoubtedly the most important aspect of any woodworking machine is its safe use. To this effect, set rules and regulations are laid down by law and must be carried out to the letter and enforced at all times. The need for such strict measures will become apparent, especially when one considers that unlike most other industries, the majority of our machines are still fed by hand, thus relying on the expertise of the skilled operator who must concentrate and exercise extreme caution at all times. A recent study by the H.S.E. has indicated that proportionally there are more serious upper body (hands, arms, torso, head) in the woodworking industry than in any other industry.

Legislation concerning all machinery and associated equipment is contained within the **P**rovision and **U**se of **W**ork **E**quipment **R**egulations 1998 (PUWER).

Since January 1993, under the Supply of Machinery (Safety) Regulations 1992, the manufacturer or the importer into the European Community has to comply with this unified European standard (signified by the **CE** mark on the machine). Or satisfy the health and safety legislation in force under PUWER. Both sets of Regulations expect new machinery to have in place approved mechanical safeguards (e.g. interlocking guards, built-in automatic braking), thus reducing the risk of accidents to operators. Machinery manufactured before 1992 was required to be updated to conform to PUWER.

The main reason for the **P**rovision and **U**se of **W**ork **E**quipment **R**egulations 1998 (PUWER) is to ensure that machinery and associated equipment does not present a risk to health and safety regardless of its age or condition. The emphasis being on the machinery having the necessary guarding, stop controls and other safety devices, that reduce the risks to the operator and other in the vicinity. The safe operation should not be reliant solely on the skills of the trained operator.

Because PUWER applies to all types of machinery, the **H**ealth and **S**afety **E**xecutive (HSE) has *published* an Approved Code Of Practice (ACoP) and guidance notes for the **H**ealth and **S**afety **C**ommission (HSC) specifically for woodworking machinery and its use. This booklet is titled 'Safe use of woodworking machinery'. *You are strongly advised to obtain and read this booklet.*

The following machines are covered in this chapter:

- Crosscutting machines
- Hand feed circular saw benches
- Dimension saws
- Panel saws
- Planing machines
- Narrow bandsaw machines
- Mortising machines
- Grinding machines

Important Note – both the Provision and Use of Work Equipment and the ACoP 'Safe use of woodworking machinery' state:

Suitability (Reg. 5)
Equipment has to be suitable for the use for which it is provided. The employer should take into account (via a risk assessment) what it is;

where it is to be used and for which purpose. If adapted or modified by the employee then they, the employee, must ensure that it is safe and suitable for its intended purpose.

Some of the following machinery may be adapted either by design or operator intention to perform operations other than "normal". For example the radial arm crosscut (9.1.2) may have design features which allow it to perform ripping, grooving and moulding operations as well as the normal 90 deg. or 45 deg. crosscutting. For operations other than "normal" the safest machines are those which have been specifically designed and are considered more suitable i.e. ripping should be carried out on circular ripsawing machine, grooving and moulding should be carried out on a spindle moulding machine (not covered in this book).

9.1   Crosscutting machines

These machines are designed to cut timber across its grain into predetermined lengths, with a straight, angled or compound-angular (angled both ways) cut. However, its primary function is to cut long lengths of timber into lengths that are more manageable.

With a standard blade and/or special cutters, some machines can also be used to carry out a variety of different operations, such as:

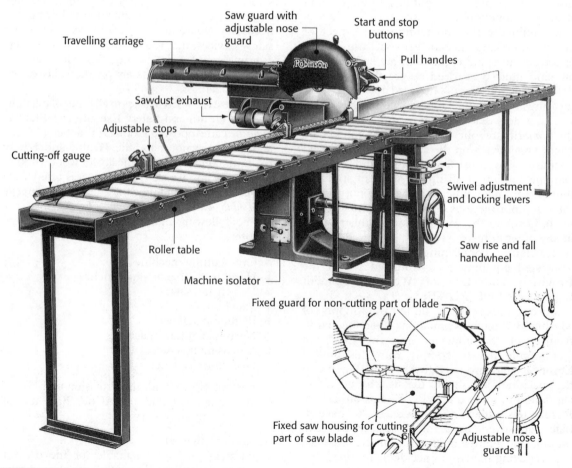

Fig 9.1   Travelling-head crosscut saw (pull-over saw)

- Trenching
- Notching

(a few examples are shown in figure 9.6)

There are two main types of crosscutting machines:

i   Travelling-head (fig. 9.1) and,
ii Radial-arm (fig. 9.2 & 9.3).

9.1.1 Travelling-head crosscut saw
(Pull-over Saw)

Figure 9.1 shows a typical saw of this type. The saw, which is driven direct from the motor, is attached to a carriage mounted on a track, which enables the whole unit to be drawn forward (using the pull handle) over the table to make its cut. The return movement is spring assisted. The length of timber to be cut is supported by a table (usually containing steel rollers) and is held against the fence. For cutting repetitive lengths,

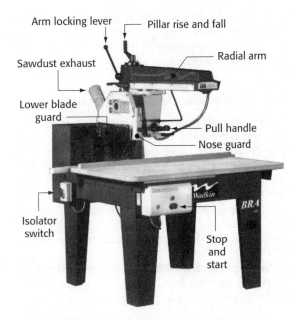

Fig 9.2   'Wadkin' radial arm crosscut saw

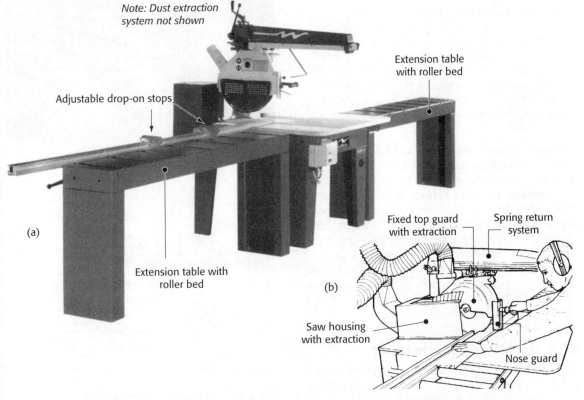

Fig 9.3   'Wadkin' radial arm crosscut saw – with extension tables

a graduated rule with adjustable foldaway stops can be provided. Angle and height adjustments are made by operating the various hand-wheels and levers illustrated.

9.1.2 Radial-arm crosscutting saw

These machines carry out similar functions to the travelling-head crosscut, but differ in their construction by being lighter. As can be seen in figure 9.2 the saw unit and carriage hangs below an arm that radiates over the table, along which it is pulled, allowing the saw to be drawn over the workpiece. The same machine is shown in figure 9.3, but in this case extension tables have been fitted to facilitate the safe cutting of long lengths of timber.

Modifying this machine to comply with PUWER and the ACoP "Safe use of woodworking machinery" would require

- Extending the saw hood side guarding to enclose as much as possible of the sawblade
- Lock-off master stop control that can be operated quickly in an emergency.
- If not already fitted, a braking device (manual or electrical) which allows the sawblade to come to rest within a maximum of ten seconds of operating the stop control.
- The alternative to above is to ensure the return spring or weighted pulley system to the head unit, once released, moves it back to a safe and housed position behind the fence line.

Some of these machines are more versatile than the travelling-head types. Not only does the carriage arm swivel 45° either way but also the saw carriage tilts from vertical to horizontal and revolves through 360°, enabling ripping, grooving and moulding operations to be undertaken. Operations other than simple 90 deg. crosscutting are not considered safe unless extra guarding and safety devices are utilized to minimise the risks e.g. anti-kickback, riving knife, chip limiting cutterblock and of course a trained operative.

9.1.3 Positioning timber to be cut

Figures 9.3 (b) and 9.4 show an operative carrying out a crosscutting operation. The length of timber must always be fully supported at both ends, to avoid tipping once the cut has been made.

Sawn ends must never be allowed to interfere with the saw blade. figure 9.5 shows how this can be avoided. 'Bowed' boards (fig. 9.5a) should be placed round side down, with their crown in contact with the table over the saw-cut line, and packing should be used to prevent the raised portion creating a see-saw effect with the possibility of it rocking on to the blade. If the board is placed round side uppermost, cut ends may drop and trap the saw blade, creating a very dangerous condition.

Similarly, when dealing with 'sprung' boards (fig. 9.5b), contact with the fence at the saw-cut line is very important, otherwise the direction of saw blade rotation could drive the cut ends of

| **Fig 9.4** | Radial arm crosscut saw in use |

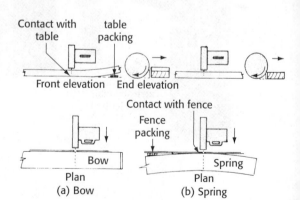

| **Fig 9.5** | Crosscutting distorted timber |

the board back towards the fence, trapping the saw blade.

Note: *Because the operative may use one hand to hold timber against the fence and the other to pull the saw, he must be constantly aware of the danger of a hand (particularly a thumb) coming in line with the saw cut.*

9.2   Hand feed circular saw benches

These are primarily used for resawing timber lengthwise in its width (ripping) or its depth (deep-cutting or deeping). They can vary the depth at which the blade projects above the saw bench table. Figure 9.7 shows a typical general purpose saw bench. The *extension* at the back of the saw bench indicates where a backing-off table must be positioned when anyone is employed to remove cut material from the delivery end (see PUWER Regulation 11 – Dangerous Parts of Machinery). The backing-off table should be no less than 1200mm when measured from the sawblade centre.

Modifying these machines to comply with PUWER and the ACoP "Safe use of woodworking machinery" would require

- Lock-off master stop control that can be operated quickly in an emergency.
- If not already fitted, a braking device (manual or electrical) which allows the sawblade to come to rest within a maximum of ten seconds of operating the stop control.

- Ensuring the position of the riving knife is no more than 8mm from the back of the sawblade at bench height. Also its curvature should conform as closely as possible to shape of the sawblade and its thickness no more than 10% thicker than the sawblade body thickness.
- Information, by way of a sign, indicating the smallest diameter sawblade allowed (i.e. 6/10ths of maximum) must be prominently displayed.

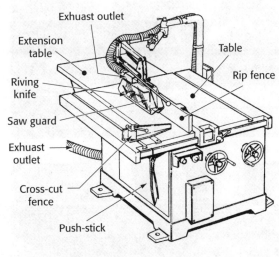

| **Fig 9.7** | Circular saw |

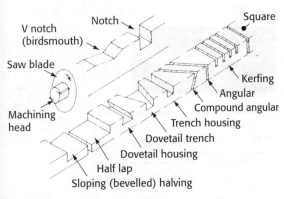

| **Fig 9.6** | The versatility of machines capable of crosscutting and trenching |

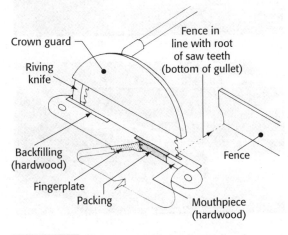

| **Fig 9.8** | Packings to a circular rip saw bench |

(a)

(b)

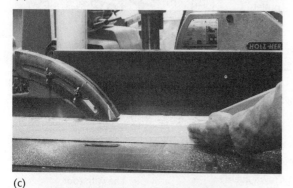

(c)

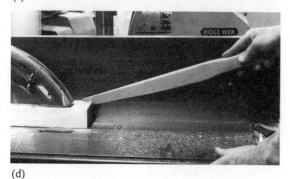

(d)

Fig 9.9   Circular saw ripping a short length of timber

Saw benches of this type use relatively large saw blades, which require packings to help prevent the saw blade deviating from its straight path. Packings are pieces of oil-soaked felt, leather or similar materials specially made by the woodcutting machinist to suit the various type of blade and their relevant position above the saw bench table. Packings, together with a hardwood mouthpiece and backfilling, can be seen in the cut-away portion of figure 9.8. Smaller diameter sawblades (e.g. 250 to 400mm dia.) may not require packings due to the less flexible cutting edge.

The mouthpiece acts as a packing stop and helps prevent the underside of sawn stock from splintering away. Like packings, it will be made to suit each size of saw blade. Backfilling protects the edges of the table from the teeth of the saw; one side is fixed to the table, the other to the fingerplate. The fingerplate lifts out the table to facilitate changing a saw blade.

9.2.1 Method of Use

Figure 9.9 shows a ripping operation being carried out. In figure 9.9a the cut is about to be started. As the cut progresses (fig. 9.9b and c), a pushstick (fig. 9.10) is used to exert pressure on the timber. During the last 300 mm of the cut only the pushstick is used (fig. 9.9d) and the left hand is moved out of harm's way. Hands are thus kept a safe distance away from the saw blade at all times. The 'birdsmouth' of the pushstick should be angled (20–30°) to suit the height of the machinist.

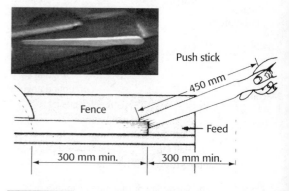

Fig 9.10   'Pushstick' and its application

By using a jig, jobs which otherwise could not be done safely are often made possible. Figure 9.11 shows a simple jig about to be used. While the cut is being made, a pushstick must be used to hold the short length of timber against the jig, as previously mentioned.

Note: A safer method of cutting wedges would be, as shown in figure 9.42(c) by using a bandsaw and purpose made jig.

For the smaller or mobile hand feed circular saw benches, one important safety item is stability. Both PUWER and the AcoP "Safe use of woodworking machinery refer as follows

● Stability of Work Equipment (Reg. 20)

Work equipment must be stable in use. This may mean the bolting down of heavy machinery; lockable castors for mobile machinery; or the addition of stabilisers/outriggers for support.

9.3   Dimension saws

These machines use a smaller saw blade, thus limiting its maximum depth of cut, depending on the size of blade and the sawbench capacity. Dimension saw benches like the one illustrated in figure 9.12 are capable of carrying out a variety of sawing operations with extreme accuracy and produce a sawn surface which almost gives the appearance of having been planed.

Modifying this machine to comply with PUWER and the ACoP "Safe use of woodworking machinery" would require

● Lock-off master stop control that can be operated quickly in an emergency.
● If not already fitted, a braking device (manual or electrical) which allows the sawblade to come to rest within a maximum of ten seconds of operating the stop control.
● Ensuring the position of the riving knife is no more than 8mm from the back of the sawblade at bench height. Also its curvature should conform as closely as possible to shape of the sawblade and its thickness no more than 10% thicker than the sawblade body thickness.
● Information, by way of a sign, indicating the smallest diameter sawblade allowed (i.e. 6/10ths of maximum) must be prominently displayed.

The main features which enable such a variety of operations to be done are:

● an adjustable double fence (tilt and length) – this adapts to suit both ripping and panel sawing

Fig 9.11   Using a jig to cut wedges. Note: a pushstick must be used to hold the wood being used as wedges against the jig

Replacement saw guard - for tilted arbor angled cutting

Detachable top saw guard for vertical sawing

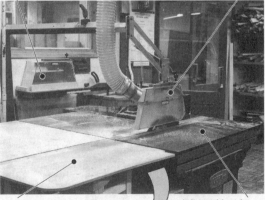

Purpose made hinged drop down extension table - shown securely locked into the horizontal position

Sliding table (shown in a locked position for a ripping operation)

Fig 9.12   Dimension saw (with sliding table)

- a cutting-off gauge – this allows straight lengths, angles or single and double mitres to be cut
- a tilting saw frame and arbor (main spindle) – this facilitates bevelled cutting, etc.
- a draw-out table – for access to the saw arbor, for saw changing
- a *sliding (rolling)* table – for panel cutting and squaring.

A smaller version of dimension saw is shown in figure 9.13.

9.3.1 Crosscutting on a Dimension Saw Bench

Figure 9.14 shows a dimension saw with a *sliding (rolling)* table being used in conjunction with a crosscutting fence set square to the blade (fig. 9.14a), set at an angle (angular) (fig. 9.14b), and by tilting the saw blade a compound angle (fig. 9.14c) cuts.

9.4   Panel saws

Figure 9.15 shows a panel saw which is similar in design to the modern dimension saw, but has features dedicated to the cutting of sheet materials. These features include:

- a smaller diameter blade – due to the limited thickness of the material to be cut

Note: options include: Extension tables and sliding table attachments

| **Fig 9.13** | Dimension saw |

- a crown guard attached to the riving knife – removing the need for a support pillar that would otherwise restrict the size of the sheet to be cut

(a) Cross-cutting square by using the cross-cutting fence and sliding (rolling) table.

(b) Cross-cutting at an angle by using an angled cross-cutting fence and sliding (rolling) table.

(c) Cross-cutting a compound angle by tilting the saw blade and using an angled cross-cutting fence and sliding (rolling) table. *Notice change of saw guard to accommodate the tilted blade.*

| **Fig 9.14** | Cross-cutting with a dimension saw |

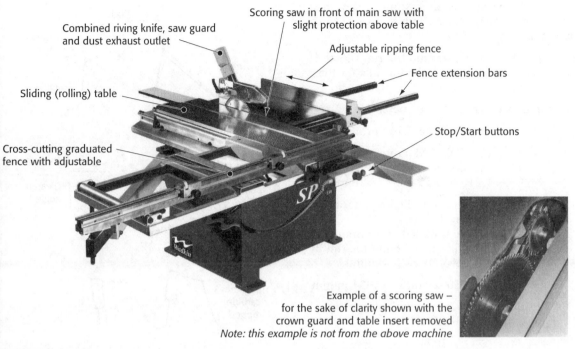

Combined riving knife, saw guard and dust exhaust outlet

Scoring saw in front of main saw with slight protection above table

Adjustable ripping fence

Fence extension bars

Sliding (rolling) table

Stop/Start buttons

Cross-cutting graduated fence with adjustable

Example of a scoring saw – for the sake of clarity shown with the crown guard and table insert removed
Note: this example is not from the above machine

Fig 9.15   'Wadkin' panel sawbench

- an extended sliding/rolling table – to cope with sheet size
- a scoring saw (optional), – a small diameter blade, pre-cutting the underside of veneered panels, thereby preventing breakout.

Modifying this machine to comply with PUWER and the ACoP "Safe use of woodworking machinery" would require

- Lock-off master stop control that can be operated quickly in an emergency.
- If not already fitted, a braking device (manual or electrical) which allows the sawblade to come to rest within a maximum of ten seconds of operating the stop control.
- Ensuring the position of the riving knife is no more than 8mm from the back of the sawblade at bench height. Also its curvature should conform as closely as possible to shape of the sawblade and its thickness no more than 10% thicker than the sawblade body thickness.
- Information, by way of a sign, indicating the smallest diameter sawblade allowed (i.e.

6/10ths of maximum) must be prominently displayed.

9.5   Saw blades

Saw blade size (diameter) in relation to the working speed of the saw spindle is very important. So much so, that every machine must display a warning notice stating the diameter of the smallest saw blade which can be fitted (under PUWER Regulation 4 – Suitability of Work Equipment, Regulation 23 – Marking, and Regulation 24 – Warning).

Saw blades are designed to suit a particular type of work and in the case of a sawbench, a rim speed (peripheral speed) of about 50m/s (metres per second) is considered suitable. Lower or higher speeds could cause the blade to become overstressed and result in a dangerous situation. Crosscut saw blades usually require a higher rim speed than ripsaw blades.

To calculate the rim speed (peripheral speed), we must know the diameter of the saw blade and its spindle speed. Let us assume a saw blade diameter of 550 mm and a spindle speed of 1750 rev/mm.

1. Find the distance around the rim (i.e. the circumference), using the formula for the circumference of a circle:

circumference = π × diameter

π may be taken to be 3.142 or 22/7

Diameter (D) = 550 mm = 0.55 m
Distance around rim = π × D
= 3.142 × 0.55 m = 1.728 m

This is the distance travelled by a tooth on the rim in one revolution.

2. Find the distance travelled every minute by multiplying the distance around the rim by the number of revolutions per minute:

1.728m × 1750 rev/mm = 3024 m/mm

3. To find the answer in metres per second (m/s), we must divide by 60:

$$\therefore \text{rim speed} = \frac{3024}{60} \text{ m/s} = 50 \text{ m/s}$$

The formula may be summarised as

rim speed (m/s) =

$$\frac{\pi \times \text{Dia of blade}}{1000} \times \frac{\text{Spindle speed (rev/mm)}}{60}$$

9.5.1 Choosing the correct saw blade

The relationship between a saw blade's size and the formation of its cutting edge will depend on one or more of the following:

- the type of sawing – crosscutting or ripping
- the type of material being cut, i.e. solid timber or manufactured boards, etc.
- the condition of the material being cut
- the finish required
- the direction of cut, i.e. across or with the grain.

Tooth shape and pitch (the distance around the circumference between teeth) greatly influence both the sawing operation and the sawn finish. Saw blades are generally divided into two groups: those most suited to cross cutting and those used for ripping and deeping (both cutting with the grain).

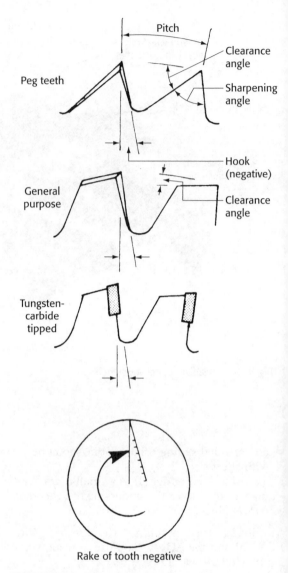

Fig 9.16   Cross-cut saw teeth

Figure 9.16 shows three different forms of tooth profile used for crosscutting. Notice that the front face of each tooth is almost in line with or sloping forward of, the radius line. This is known as 'negative hook' and produces a clean cut.

Figure 9.17 shows how each tooth of a ripsaw blade slopes back from the radius line to produce a true hook shape, termed 'positive hook'. It is this shape which enables the tooth to cut with a riving or chopping action. The large gullet helps keep the kerf free of sawdust.

A tungsten carbide tipped (TCT) tooth is

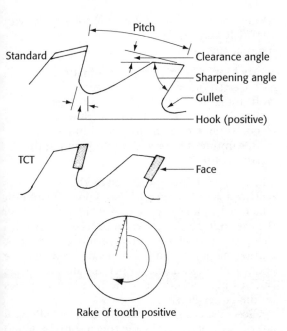

Rake of tooth positive

Fig 9.17 Rip-saw teeth

ideal for cutting hard or abrasive material because its hardness means it stays sharp much longer than the conventional saw tooth. However, its eventual re-sharpening does involve the use of special equipment and, because of this, the maker of the saw blade or an appointed agent usually undertakes the task.

Knowing the number of teeth and the diameter of the saw blade will enable the tooth pitch to be determined:

tooth pitch =

$$\frac{\text{rim distance (circumference of saw blade)}}{\text{number of teeth around the rim}}$$

where rim distance = $\pi \times$ diameter

For example, if a saw blade has a diameter of 500 mm and 80 teeth:

$$\text{Tooth pitch} = \pi \times \frac{\text{Diameter}}{\text{number of teeth}}$$

$$= \frac{3.142 \times 500 \text{ mm}}{80}$$

$$= 19.6 \text{ mm}$$

Note: In general, the shorter the pitch, the finer the cut.

With the exception of certain saw blades which are designed to divide timber with the least possible waste, the kerf left by the saw cut

must be wide enough not to trap the saw blade. Figure 9.18 shows how this is achieved. The hollow-ground saw blade (fig. 9.18a), which produces a fine finish and is used for crosscutting, uses its reduced blade thickness, whereas the parallel-plate saw blade (fig. 9.18b) uses its spring set (teeth bent alternately left and right). TCT (**T**ungsten **C**arbide **T**ipped) saw blades (fig. 9.18c) rely upon the tip being slightly wider than the blade plate to provide adequate plate clearance.

Figure 9.18 also shows a saw blade mounting. The holes in the saw blade must match the

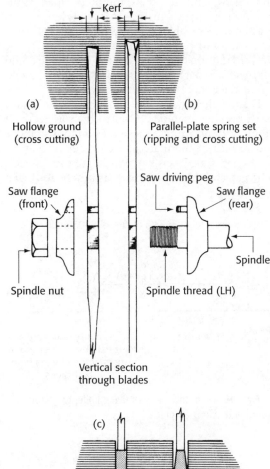

Vertical section through blades

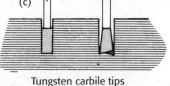

Tungsten carbile tips
(ripping and crosscutting)

Fig 9.18 Circular saw blades and mounting to a sawbench spindle

size and location of both the spindle and the driving peg, which is incorporated in the rear flange. The front flange is positioned over the peg before the spindle nut is tightened.

9.5.2 Guarding

Figure 9.19 shows how the riving knife should be positioned in relation to the size of the saw blade. Its thickness must exceed (by approximately 10%) that of the blade if a parallel-plate saw blade is used. However, its thickness should be less than the 'kerf' (fig 9.18). The purpose of the riving knife is to act as a guard at the exposed rear of the saw blade and to keep the kerf open, thus attempting to prevent the timber closing (binding) on the saw blade. If the saw cut was allowed to close (with case-hardened timber, etc.), the upward motion of the back of the saw blade could lift the sawn material and possibly project it backwards (kickback) towards the operator.

During the sawing process, the crown guard must provide adequate cover to the saw teeth. It should extend from the riving knife to just above the surface of the material being cut (it may be necessary to use a crown guard extension piece). The gap should be as narrow as practicable, (fig. 9.19).

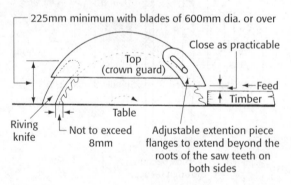

Fig 9.19 Arrangement of riving knife, top (crown) guard and extension piece

9.6 Planing machines

Planing machines generally fall into three groups:

1   hand-feed surface planers
2   thicknesser or panel planers
3   combined hand – and power-feed planers.

Modifying the above machinery to comply with PUWER and the ACoP "Safe use of woodworking machinery" would require

- Lock-off master stop control that can be operated quickly in an emergency.
- If not already fitted, a braking device (manual or electrical) which allows the cutterblock to come to rest within a maximum of ten seconds of operating the stop control.

The quality of finish produced by these machines will to a large extent depend upon the rate at which the timber is passed over the cutter or, in the case of the thicknesser, under the cutter. Also the speed at which the cutters revolve around their cutting circle (cutting periphery) – on average about 1800 m/min (metres per minute), or 30 m/s (metres per second), and the number of blades on the cutter block.

Close inspection of a planed surface will reveal a series of ripples left by the rotary cutting action of the blades. These marks may have a pitch of 1–3 mm. As shown in Figure 9.20, the shorter the pitch, the smoother the surface finish; there-

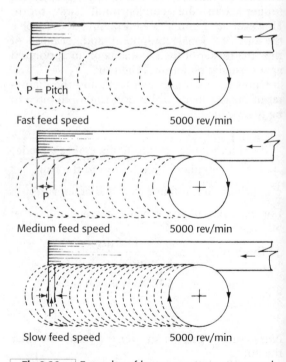

Fig 9.20 Examples of how a constant cutter speed can produce different surface finishes with different feed speeds

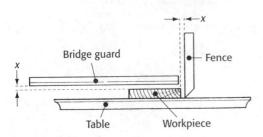

(a) Adjutment of the bridge guard

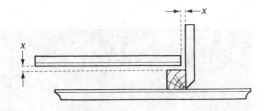

(b) Edging

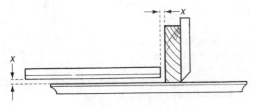

(c) Flatting and edging rectangual stock

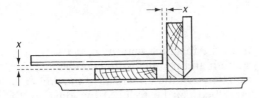

(d) Flatting and edging small square stock

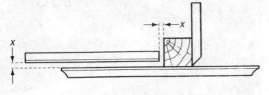

(e) Flatting and edging large square stock

Note: Bridge guard to be adjusted as close to the workpiece and fence as possible (x)

Fig 9.26   Positioning the bridge guard for flatting and edging operations

Figure 9.26 shows how the bridge guard should be positioned when carrying out the various surfacing operations. Figure 9.26(a) shows the position for flatting, figure 9.26(b) that for edging and figure 9.26(c) for when flatting and edging are carried out one after the other. Figure 9.26(d) & (e) show flatting and edging squared stock.

Hand positions for flatting are shown in figure 9.27 – hands must never be positioned over the cutter. Figure 9.27(a) shows the approach position. Once the timber has passed under the bridge guard, the left hand is repositioned on the delivery side of the table (fig. 9.26(b)). As the process continues, the right hand follows (fig. 9.26(c)), the timber being pressed down on to the table during the whole operation. The entire process is repeated until the desired flatness is obtained.

In figure 9.28 edging is being undertaken with the bridge guard set as shown in figure 9.26(c), again, the process is repeated until the edge is both straight and square to its face side.

When dealing with slightly bowed or sprung timber (badly distorted timber should be shortened or sawn straight, using a jig if necessary), it should be positioned fully on the infeed table (round side/edge up or hollow side/edge down – fig. 9.29), and then passed over the cutters by making a series of through passes until straightened.

Where, for reasons of safety short pieces of timber cannot be planed in accordance with figure 9.26(a), a pushblock (fig. 9.30) offering a firm and safe handhold should be used. figure 9.31 shows a pushblock in use.

Wherever possible, the direction of wood grain should run with the cutters (fig. 9.32). In this way, tearing of the grain is avoided and a better surface finish is obtained.

9.6.6 Thicknessing

This process involves pushing a piece of timber, face down, into the infeed end of the machine, where it will be met and gripped by the fluted (serrated) infeed roller and mechanically driven under the cutters.

Depending on the size and make of machines and the thickness scale setting, the cutters could remove up to 3 mm from the thickness of the timber, after which it is steadied by the outfeed roller, (a smooth-surfaced roller, so as not to bruise the surface of the wood) and delivered

(a) Approach from the infeed table

(b) Repositioning towards the outfeed table

(c) Follow through on the outfeed table to delivery end

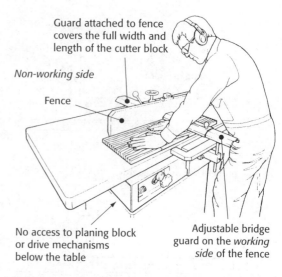

Guard attached to fence covers the full width and length of the cutter block

Non-working side

Fence

No access to planing block or drive mechanisms below the table

Adjustable bridge guard on the *working side* of the fence

(d) Basic safeguarding requirements

Fig 9.27   The process of 'flatting timber before edging

from the machine with its upper surface planed smooth and parallel with its underside. Any friction between the underside of the timber and the table bed can be reduced by adjusting the two antifriction rollers to suit the condition of the

Notice the easy access to the control panel

Fig 9.28   'Edging' after 'flatting'

9.7.5 Use

During all sawing operations, the top guard must be set as close to the workpiece as practicable. This not only protects the operative from the saw blade but also provides the blade with maximum support via the guides and the thrust wheel or thrust roller assembly (fig. 9.39).

Figure 9.42 shows, with the aid of a ripping fence.

a) a ripping operation – towards the end of the cut, the pushstick will be used to push the side nearest the fence.
b) deep cutting
c) wedge cutting

When cutting into a corner (fig. 9.43), short straight cuts are made first (this also applies when making curved freehand cuts; fig. 9.44). An exception to this rule (unless a jig was used) would be when removing waste wood from a haunch (fig. 9.45), in which case the cuts at 'A' with the grain are made first, to reduce the risk of cutting into the tenon, then the cuts at 'B'. The small portion left at 'C' would, depending on the blade width, probably have to he nibbled away by making a series of short straight cuts.

The operative in figure 9.46 is using a bandsaw to cut curves freehand.

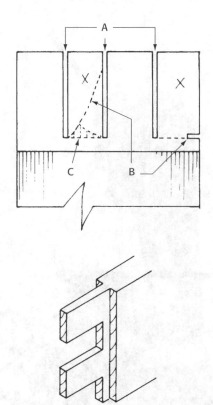

9.8   Mortising machines

These machines cut square sided holes or slots to accommodate a tenon. The hole is made

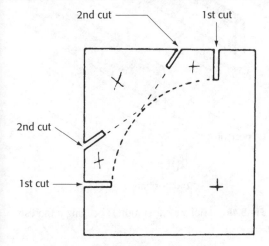

Fig 9.44   Making a curved cut into a corner

Fig 9.45   Removing waste wood from a haunched tenon

Fig 9.46   Cutting a curve freehand

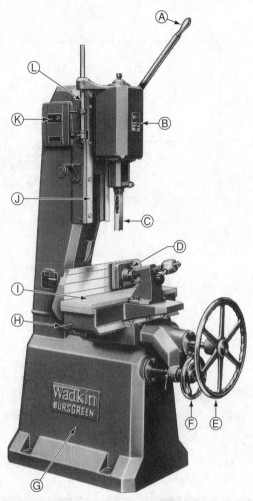

Fig 9.47 'Wadkin' hollow chisel mortiser. A – Operating levers; B – Mortising head; C – Hollow chisel and auger; D – Clamp (should be faced with a wooden plate); E – Handwheel; operates table longitudinal movement; F – Handwheel; operates table cross-traverse; G – Main frame; H – Table stop bar; I – Work table (timber packing) and rear face; J – Mortising head slideway; K – Control; start and stop buttons; L – Depth stop bar (mortise depth adjustment)

either by a revolving auger bit inside a square tabular chisel, or by an endless chain with cutters on the outer edge of each link. Machines are made to accommodate either method or a combination of both. Namely:

- Hollow Chisel Mortiser
- Chain Mortiser★
- Combined Chain and Chisel mortiser★

★ *The latter two machines are not covered by this book.*

9.8.1 Hollow chisel mortiser

Figure 9.47 shows the various components of this machine (see also fig. 6.12). As the mortising head is lowered, the auger bores a hole while the chisel pares it square. The chippings are ejected from slots in the chisel side. After reaching the required depth, which for through mortise holes is about two-thirds the depth of the material, the procedure is repeated along the desired length of the mortise. At all times during the operation the workpiece must be held secure with the clamp against the table and rear face (fence). The workpiece is then turned over and reversed end for end, to keep its face side against the fence, and the process is repeated to produce a through mortise (see fig. 9.48).

In order to cut the square hole cleanly and safely without damaging the cutters or 'blueing' them (as a result of overheating due to friction), a gap of 1 mm must be left between the spur edges of the auger bit and the inside cutting edges of the square chisel, as shown in figure

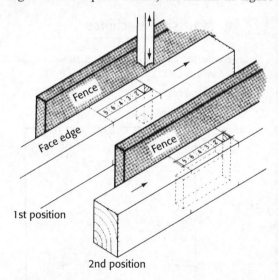

Fig 9.48 Hollow chisel mortiser cutting a mortise hole

employer to ensure only adequately trained personnel are authorised to operate machinery and equipment. The specific reference is:

- *Specific risks*
  Were the use of equipment has a known risk then, in such cases, the employer must ensure its use is restricted to trained persons.
- *Employee information, instruction & training*
  Employers must ensure that adequate and sufficient information, instruction and training is given to employees regarding the use of work equipment and the environment involved.

It is important, therefore, that all trainees make an in-depth study of all aspects of woodworking machinery safety. Itemised below are areas of study together with some of their relevant safety factors.

9.12.1 Provision and use of Work Equipment Regulations 1998

These regulations apply to all machines, not just those utilised in the timber/construction industries. They are therefore of general legislative application. However, they are legal requirements and should be understood by all who purchase and/or use woodworking machinery. Guidelines in the form of an Approved Code of Practice (ACoP) 'Safe use of woodworking machinery' should be read to help illustrate and interpret these regulations.

9.12.2 Noise at Work Regulations 1989

The more traditional woodworking machines illustrated in this chapter are typically found in the joinery workshop. Inherently, if not modified, they are noisy in use and in many cases exceed the noise levels considered 'acceptable' according to these regulations (i.e. below 85dB(A)). The regulations state that noise measurements should be taken and recorded. Suitable safeguards to the health, safety and welfare of employees must be taken by employers if measurements exceed certain criteria.

9.12.3 Physical condition of the machine and its attachments

Know the location and operation of:

- the electricity isolating switch (PUWER – must be labeled to identify machine)

- the machine controls (PUWER – Stop control must be prominently positioned)
- guards and their adjustment.

9.12.4 Setting-up and guarding the machine

- Know how to isolate the machine.
- Ensure that the blade/cutters are of the correct type, size and shape and are sharp.
- Ensure that all adjustment levers are locked securely.
- Check that all guards and safety devices are in place and secure.
- Ensure that all the adjusting tools have been returned to their 'keep'.
- Make sure that pushsticks and blocks are at hand.
- Ensure that work and floor areas are free from obstruction.

9.12.5 Suitability of the material to be cut

- Understand the cutting characteristics of the material.
- Check length and section limitations.

9.12.6 Machine use

- Always allow the blade/cutters to reach maximum speed before making a cut.
- Know the correct stance and posture for the operative.
- Use an assistant as necessary.
- Never make guard adjustments until all moving parts are stationary and the machine has been isolated.
- Never make fence adjustments when a blade/cutter is in motion.
- Never make fence adjustments within the area around the blade/cutters or other moving parts until those parts are stationary and the machine has been isolated.
- Concentrate on the job – never become distracted while the machine is in motion.
- Never allow hands to travel near or over a blade/cutter while it is in motion.
- Never leave a machine until the blade/cutters are stationary
- Always isolate the machine after use.

9.12.7 Personal safety

- Ensure that dress and hair cannot become caught in moving parts or obstruct vision.
- Finger rings should never be worn in a

machine shop, for fear of directing splinters of wood into the hand or crushing the finger if the hand becomes trapped.

● Footwear should be sound with non-slip soles of adequate thickness and firm uppers to afford good toe protection.

● Wear eye protection.

● Wear ear protection.

Basic Woodworking Joints

10

There are many different joints that the carpenter and joiner may use. This chapter is concerned mainly with those made by hand.

Joints generally fall into three categories and carry out the following functions:

Category	Joint	Function
a Lengthening	End	To increase the effective length of timber.
b Widening	Edge	To increase the width of timber or board material.
c Framing	Angle	To terminate or to change direction

10.1 Lengthening – end joints (fig. 10.1)

Where timber is not long enough, a suitable end joint must be made. The type of joint used will depend on the situation and end use.

10.1.1 Lap Joint (fig 10.1a)

The two adjoining lengths of timber are lapped at their ends.

10.1.2 Butt joint (cleated)

Figure 10.1(b) shows two methods of securing an end butt joint. Depending on whether the joint is to be concealed or not, and the required strength, a *'single'* or *'double'* cleat can be used.

10.1.3 Scarf joint

Figure 10.1(c) shows two methods of making a scarf joint. The first, for structural use will require a slope of 1 in 12 or less. The second method incorporates a hook that enables the joint to be tightened with folding wedges.

10.1.4 Laminated joint

By laminating (overlapping) different lengths of timber together with nails and/or glue, large long lengths of timber can be manufactured. Basic principles are shown in figure 10.1(d).

10.1.5 Finger joint

A finger joint is shown in figure 10.1(e). This is produced by machine, then glued and assembled by controlled end pressure. This is a useful method of using up short ends and upgrading timber – after the degraded portion or portions have been removed, the remaining pieces are rejoined.

10.1.6 Half-lap & sloping halving joints

These joints are also used as 'Framing Joints' see figure 10.11 and section 10.3.2.

They can also be used for lengthening and intersecting wallplates in floor and roof construction.

10.2 Widening – edge joints

Whether the joint is to be permanent (glued), or flexible (dry joint – without the use of an adhesive.), will depend on its location.

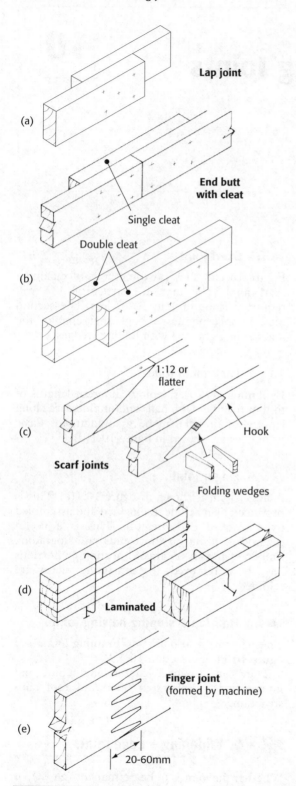

(a) **Lap joint**

(b) **End butt with cleat**

Single cleat

Double cleat

(c) 1:12 or flatter

Hook

Scarf joints

Folding wedges

(d) **Laminated**

(e) **Finger joint** (formed by machine)

20-60mm

Fig 10.1 Lengthening joints (see also half lap & sloping halving joints – fig 10.11)

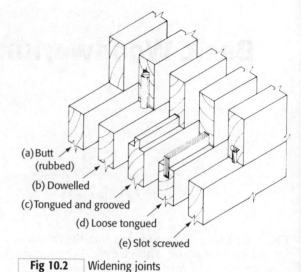

(a) Butt (rubbed)

(b) Dowelled

(c) Tongued and grooved

(d) Loose tongued

(e) Slot screwed

Fig 10.2 Widening joints

10.2.1 Edge glued joints

All edge glued joints enable a board's width to be increased, examples of these joints are shown in figure 10.2 which include:

- butt joints
- dowelled joints
- tongued joints
- loose tongued joints
- slot-screwed joints

Whatever method of jointing is chosen, it is always wise to try to visualise how a board will react if subjected to moisture. As stated in section 1.7.3 and shown in figures 1.36 & 1.53, tangential-sawn boards are liable to 'cup'. Figure 10.3 shows a way to minimise this effect.

a **Butt joint** (fig.10.2a) – this is the simplest of all edge joints and is the basis of all the other forms shown in figure 10.2. If the joint is to be glued, it is important that the adjoining edges match perfectly. Figure 10.5 shows how this is achieved:

1. The boards are first marked in pairs (fig. 10.5(a)).
2. Each pair is planed straight and square by using a plane with the appropriate sole length (preferably long-soled try-plane (fig. 10.5(b)).
3. They are then repositioned edge to edge to check that no light shows through the joint and both faces are in line (fig. 10.5(c)).
4. Glue is applied while both edges are positioned as if they were hinged open. They are then turned edge to edge and

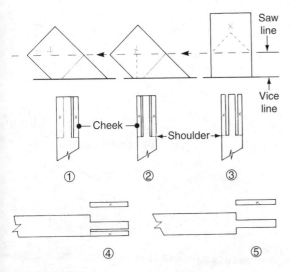

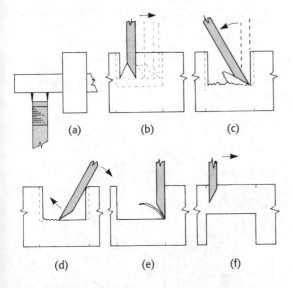

Fig 10.14 | Sequence of cutting a tenon

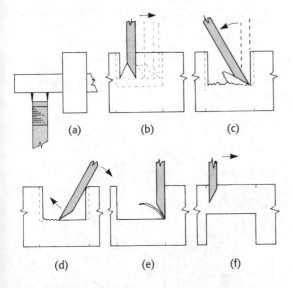

left, chop and gently lever waste wood into the hole formed.

d Repeat until midway into the workpiece, carefully lifting out chippings at each level. (*Note how the waste left on the ends of the hole prevents damage to the mortise hole during this process*).

e Square the ends of the mortise hole.

f Turn the workpiece over and repeat the process, as shown in figure 5.64 the bench should be protected at this stage if a through mortise is to be cut.

10.3.4 Bridle joints (fig 10.16)

Except for the '*corner*' bridle – also known as an *open* or *slot* mortise – bridle joints slot over through-running members. Named examples are shown in figure 10.16. They are cut in a similar manner to tenons and halving joints.

Fig 10.15 | Chopping a mortise hole (sectional details) – back lines of mortise hole omitted for the sake of clarity

a Set the mortise gauge to the width of the chisel, which should be as near as possible to one third the width of the material being mortised.

b After having secured the workpiece (see figure 5.64), chop a hole approximately 15 to 20 mm deep and 4 mm in from one end.

c Working from left to right or from right to

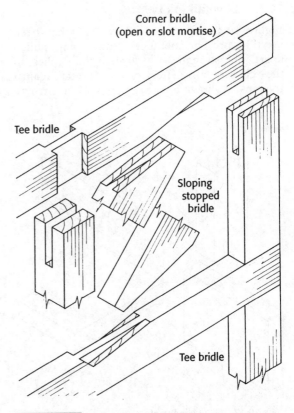

Fig 10.16 | Bridle joints

10.3.5 Dowelled joints (fig 10.17)

Useful alternatives to mortise-and-tenon joints for joining members in their thickness (fig. 10.17(a)) or as a means of framing members in their width (fig. 10.17(b)). Dowel and hole preparation is similar to the methods shown in figure 10.6. Correct alignment of dowel with hole is critical, but this problem can be overcome by using a dowelling jig (fig 5.95 a template and guide for boring holes accurately).

10.3.6 Notched and cogged joints (fig 10.18)

The notches of a *notched joint* are used to locate members in one or both directions and as a means of making any necessary depth adjustments (joist to wallplates etc.). *Cogged joints* perform a similar function, but less wood is removed, therefore generally leaving a stronger joint – they do, however, take much longer to make.

10.3.7 Dovetail joints

Figure 10.19 shows how dovetailing has been used to prevent members from being pulled apart. The strength of a dovetail joint relies on the self-tightening effect of the dovetail against the pins, as shown by the direction of the arrows.

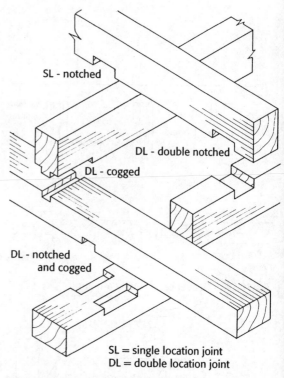

Fig 10.18 Notched and cogged joints (exploded isometric views)

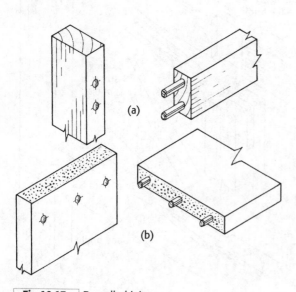

Fig 10.17 Dowelled joints

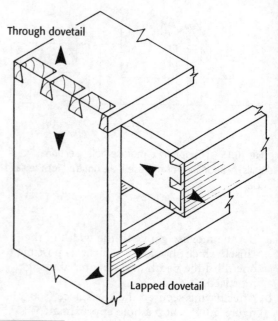

Fig 10.19 Dovetail joints

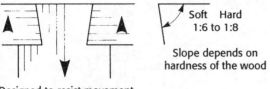

Designed to resist movement
in the direction of the arrows

Soft   Hard
1:6 to 1:8

Slope depends on
hardness of the wood

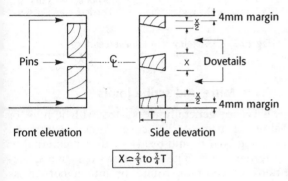

Pins →

Dovetails

x

4mm margin

4mm margin

Front elevation   Side elevation

$$X \simeq \frac{2}{3} \text{ to } \frac{3}{4} T$$

N.B. GUIDE ONLY

Fig 10.20   Proportioning a dovetail joint

Dovetail slopes can vary between 1 in 6 to 1 in 8, depending on the physical hardness of the wood. The number of dovetails and their size will vary with the width of the board. The dovetails are usually larger than the pins (except those produced by machine – which are of equal width), and a guide to their proportions is given in figure 10.20.

Figure 10.21 shows a method of marking and cutting a single through dovetail;

a Set the marking gauge to the material thickness.
b Temporarily pin together those sides (in pairs) which are to be dovetailed at their ends – gauge all round.
c Using a bevel or dovetail template, mark the dovetails (see fig. 5.12).
d Using a dovetail saw cut down the waste side to the shoulder line.
e Cut along the shoulder line and remove the cheeks.
f Divide the sides and mark off the pins.
g Cut down to the shoulder line with a dovetail saw.
h Remove waste with a coping saw, then pare square with a chisel.

The joint should fit together without any further adjustments!

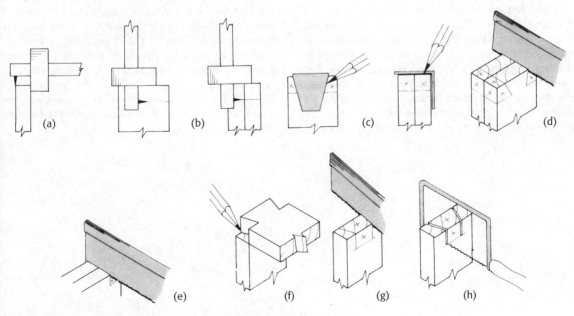

Fig 10.21   Marking-out and cutting a dovetail joint

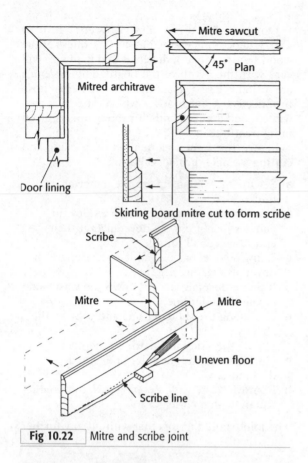

Mitred architrave

Door lining

Mitre sawcut

45° Plan

Skirting board mitre cut to form scribe

Scribe

Mitre

Mitre

Uneven floor

Scribe line

Fig 10.22   Mitre and scribe joint

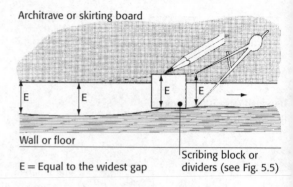

Architrave or skirting board

Wall or floor

E = Equal to the widest gap

Scribing block or dividers (see Fig. 5.5)

Fig 10.23   Scribing to an uneven surface

Some joiners prefer to cut the pins first and the dovetails last. Both methods are acceptable, but in my opinion the sequence described is quicker and tends to be more accurate. Probably the first opportunity the student gets to use this joint is during the construction of a toolbox, chest or case – dovetail joints should be the first choice when constructing one or more units of the *Porterbox* projects (situated after section 12).

10.3.8 Mitre and scribe joints

These are generally associated with joining or abutting trims – i.e. cover laths, architraves, skirting boards, and beads – either at external or internal angles. The joint allows the shaped sections to continue round or into a corner, as shown in figure 10.22.

A mitre is formed by bisecting the angle formed by two intersecting members and making two complementary cuts. The scribe joint has its abutting end shaped to its own section profile, brought about by first cutting a mitre (figure 10.22).

Note: The scribe joint used should not be confused with a surface scribe, as this is where a joint has to be made between a trim and an uneven surface, such as a floor, wall, or ceiling, scribing in this case provides a means of closing any gaps. Figures 10.22 and 10.23 show that, by running a gauge line parallel to the uneven surface, an identical contour will be produced (see also Carpentry & Joinery Book 3 (wall trims and finishes)).

Wood Adhesives

An adhesive is a medium that allows the surfaces of two or more items to be attached or bonded together. Adhesives are made from either natural or synthetic (man-made) materials. They come in a liquid (one or two part), powder form, a combination of both powder and liquid, or as a semisolid which requires melting.

Table 11.1 lists nine adhesive types together with their classification, main characteristics, moisture resistance, any gap-filling properties and their general usage. The following notes briefly describe these adhesives and some of the technology associated with them.

11.1 Adhesive types

11.1.1 Casein

Derived from dairy by-products that are dried, treated and mixed with chemical additives to produce a powder which, when mixed with water, is ready for use. It is used in general joinery assembly work and in the manufacture of plywood. It tends to stain some hardwoods.

11.1.2 Urea Formaldehyde (UF)

A very widely used synthetic resin adhesive, for such things as general assembly work and binding within some manufactured boards (including MDF). Strong mixes can achieve moisture-resistant (MR) requirements. Most are designed for close contact jointing, but formulations are available to satisfy gap-filling requirements (see section 11.2.3).

When set (chemically cured) these adhesives will either be clear or lightly coloured so glue lines can be concealed.

11.1.3 Melamine Urea Formaldehyde (MUF)

An adhesive with more melamine than urea, which can enhance moisture resistance to bring it above boil-resistant (BR) rating. Its usage is similar to that of UF.

11.1.4 Phenol Formaldehyde (PF)

The main purpose of these adhesive types is to provide the best moisture resistance (weather and boil proof – WBP) to structural plywood. They are also used as a binder in particleboard and wafer boards. Glue lines may show as slight red/brown staining to the wood.

11.1.5 Resorcinol Formaldehyde (RF)

Because of their WBP properties these adhesives are highly suited to the assembly of external structures and marine applications. They are, however, expensive. Glue lines can result in the staining of wood. Again, the glue lines may show a red/brown staining to the wood.

11.1.6 Polyvinyl Acetate (PVAc)

Thermoplastic adhesives consisting of a simple to use, one-part water-based emulsion, with additives to produce either a standard interior type, cured mainly by evaporation and used extensively for glueing internal joinery components and veneering. Or, an improved PVAc, which gives higher moisture resistance by inducing a chemical reaction. This so-called 'cross-linking' will put this type of PVAc in the class of a thermosetting adhesive.

Table 11.1 Adhesives characteristics (general guide only as properties can differ)

Adhesive classification			Moisture resistance	Class	Gap filling	General usage
N	Casein		Poor	Int	Yes	Not in general use today – has been used for assembly work and interior plywood
S	Urea formaledehyde (UF)	Amino-plastic	Thermo setting*	Fair	MR	Some types	Probably the most common adhesive – general purpose assembly work, manufactured boards (MDF) and veneering
	Melamine urea formaldehyde (MUF)			Good	BR		Moisture resistant properties of UF can be improved by this additive – MUF is used in some particleboards and plywood
	Phenol formaldehyde (PF)	Phenolic		Very good	WBP	Yes	Structural plywood and binder in particleboards and wafer boards – not often for joints
	Resorcinol formaldehyde (RF)			Very good	WBP	Yes	Outdoor timber structures – but not often used on its own because of its high cost RF may be linked with PF to produce PRF
	Polyvinyl acetate (PVAc)		Thermo Plastic	Poor to good	Int or MR†	May be	Internal assembly of joinery items and veneering (a general purpose adhesive group)
S&N	Contact		Solvent or emulsion based	Fair	Int		Veneers of wood and plastic laminate
	Hot melt		Thermo-plastic		Int		Edge veneering and spot or strip glueing (glue gun) – bond strength is relatively low
S	Epoxy	Thermo setting			Int	Yes	Very limited to specialist use – usually formulated to enable wood to be bonded to metal and glass reinforced plastics (GRP)

N – adhesives in the main are of natural origin; S – synthetic resins; WBP – weather and boil proof; BR – boil resistant; MR – moisture resistant; INT – interior;

* Note: All are two part (component – powder & liquids forms available) adhesives; † Special formulation for exterior use, depending on formulation may be classed as a thermosetting adhesive

Thermoplastics – Thermoplastic adhesives may be in the form of fusible solids which soften by heat, or in a soluble form. Generally after heating thermoplastics will regain their original form and degree of strength

Thermosets – Thermosetting resins become solid either by a chemical reaction, or via a heat source; however, once set they, unlike thermoplastics, can not be reconstituted by heat

11.1.7 Contact adhesives

Made of natural or synthetic rubber and a solvent that evaporates when exposed to the air, giving off a heavy flammable vapour. They are used in the bonding of wood and plastic veneers (laminated plastics) to wood based materials. Bonding is achieved by coating both surfaces to be joined, leaving them to become tacky (for a time specified by the manufacturer) and then laying one onto the other while excluding any air. Bonding is instantaneous on contact (hence the name *'contact adhesive')*, with the exception of *thixotropic* types which allow a certain amount of movement for minor adjustments.

Note: Some generally available contact adhesives are emulsion based, non-flammable and less hazardous.

11.1.8 Hot melt synthetic (thermoplastic) adhesives

Are either semisolid rods or pellets or come in the form of tapes or films, which are melted for application by heat. As they cool, they re-harden to their original strength. The most common type of application is with an edge-banding machine, used for applying veneers of wood, plastics, and plastic laminates. Handheld glue guns are quite popular for spot glueing or running narrow joints; glue guns use cylindrical adhesive 'slugs', which are available in various formulations.

11.1.9 Epoxy synthetic (thermosetting) adhesives

This type of adhesive is limited to specialised applications, for example, glueing wood to metal or glass reinforced plastics (GRP) and small repair work.

11.2 Adhesive characteristics

11.2.1 Form

Adhesives may be of the 'one' or 'two' component type, liquid, powder or both. Two component types become usable either by applying them direct from the container, mixing the components together, or by applying each part separately to the surfaces being joined. Some types, however, have to be mixed with water.

11.2.2 Moisture resistance

Refers to the adhesive's inherent ability to resist decomposition by moisture. Resistance is classified as follows:

- INT (interior) – joints made with these adhesives will be resistant to breakdown by cold water
- MR, moisture and moderately weather-resistant joints will resist full exposure to weather for a few years when made with these adhesive types.
- BR (boil resistant) – joints made with these adhesives will resist weather but fail when subjected to prolonged exposure. They will, however, resist breakdown when subjected to boiling water.
- WBP (weather and boil proof) – joints made from these adhesives will have high resistance to weather, micro-organisms, cold and boiling water, steam and dry heat.

Reference to the above categories is also made in table 4.1 with regard to the bonding of plywood veneers; these are British Standard Terms.

11.2.3 Gap filling

Adhesives should be capable of spanning a 1–1.3 mm gap without crazing. They are used in situations where a tight fit cannot be assured and where structural components are to be bonded.

11.2.4 Bond pressure

Refers to the pressure necessary to ensure a suitable bond between two or more surfaces joined together. Pressure may be applied simply by hand, as with contact adhesives. With items of joinery, mechanical, manual presses or clamps (see section 5.11.1) of various shapes and sizes are usually employed. Wood wedges can be used not only to apply pressure but also to retain a joint permanently. The length of time needed to secure a bond will vary with each type of adhesive, its condition and the surrounding temperature.

11.2.5 Assembly (closed) time

A period of time will be required to accurately set cramping devices and correctly position frame components; each type and make of adhesive will state the time allowed for this.

11.2.6 Storage or shelf life

Period of time that a containerised adhesive will remain suitable for use. Beyond this period, marked deterioration may occur, affecting the strength and setting qualities.

Once the adhesive's components have been exposed to the atmosphere, the storage life may well be shortened. 'Shelf life' therefore may or may not refer to the usable period after the initial opening, always note the manufacturer's recommendations.

11.2.7 Pot life

Period of time allowed for use after either mixing or preparing the adhesive.

11.3   Application of adhesives

Methods and equipment used to apply adhesives will depend on the following factors:

- type of adhesive
- width of surface to be covered
- total surface area
- work situation
- clamping facilities available
- quality of work.

The spreading equipment could be any one of the devices listed below:

- mechanical spreader
- roller
- brush
- spatula
- toothed scraper.

11.4   Safety precautions

All forms of adhesives should be regarded as potentially hazardous if they are not used in accordance with the manufacturer's instructions, either displayed on the container or issued as separate information sheet.

Depending on the type of adhesive being used, failure to carry out the precautions thought necessary by the manufacturer could result in:

- an explosion – due to the adhesives flammable nature or the flammable vapour given off by them
- poisoning – due to inhaling toxic fumes or powdered components
- skin disorders – due to contact while mixing or handling uncured adhesives. Always cover skin abrasions before starting work.

Where there is a risk of dermatitis, use a barrier cream or disposable protective gloves. Always wash hands thoroughly with soap and water at the end of a working period.

Finally, the manufacturer's instructions on handling precautions should cover the following:

- good 'housekeeping'
- skin contact
- ingestion
- eye protection
- fire risk
- toxicity.

Fixing Devices

The decision as to how a piece of timber is fixed and which device to use is usually left to the joiner – unless the designer states otherwise.

Where the method of jointing and fixing are critical to the stability of the structure and or its attachments, a structural engineer must be responsible for its design, in which case the specification must be followed to the letter.

In general terms where ever fixing devices are to be used, the following factor should be kept in mind;

- location,
- strength requirements,
- resistance to corrosion,
- appearance,
- availability,
- cost.

Fixing devices can be broadly classified as:

- Nails
- Wood screws
- Bolts (threaded)
- Metal fixing plates
- Plugs
- Combination plugs
- Cavity fixings
- Anchor bolts
- Chemical fixings

12.1   Nails

Nails offer the quickest, simplest, and least expensive method of forming or securing a joint and, provided the material being fixed is suitable, the nails are the correct type, size (length, and gauge), and correctly positioned to avoid splitting, a satisfactory joint can be made.

12.1.1 Nail selection

Table 12.1 not only illustrates the different types of nails, pins, and staples but also gives examples of the application.

Whenever a nailed joint forms part of a structural component, the building designer should specify its size (which takes into account both its length and gauge).

However, as a general guide (as shown in fig. 12.1(a)), when fixing timber to timber the nail length should be $2\frac{1}{2}$ times the thickness of the timber being fixed. One exception to this rule is, where as shown in figure 12.1(b), through nailing has been specified, such as you may find used on a battened door (see book 3).

12.1.2 Nail holding power

When dismantling a joint or fixture which has been nailed, pay particular attention to the effort

Key to Tables 12.1, 12.2, 12.6 and 12.7

Materials
A – aluminium alloy   B – brass   BR – bronze
C – copper   P – plastics   S – steel
SS – stainless steel

Finish/treatment
B – brass   BR – bronze   CP – chromium
G – galvanised   J – jappanned (black)   N – nickel
SC – self-coloured   SH – sherardised   Z – zinc
BZ – bright zinc

Head shape
CKS – countersunk   DM – dome   RND – round head
RSD – raised head   SQ – square

Drive mechanism
SD – Superdriv (Pozidriver)   SL – slotted (screwdriver)
SP – square head (spanner)
(Note: *Superdriv* is the successor to *Pozidriv*.)

Table 12.1   Nails, pins, and staples

Nail type	Material	Finish or treatment	Shape or style	Application
Wire nails				
Round plain-head	S	SC		Carpentry; carcase construction; wood to wood
Clout (various sized heads)	S, C	SC, G		Thin sheet materials; plasterboard; slates; tiles; roofing felt
Round lost-head	S	SC		Joinery; flooring; second fixing. (Small head can easily be concealed.)
Oval brad-head	S	SC		Joinery; general-purpose. Less inclined to split grain.
Oval lost-head	S	SC		As for Oval brad-head.
Improved nails				
Twisted shank	S	SC, G		Roof covering; corrugated and flat materials, metal plates, etc.; Flooring; sheet materials. Resist popping (lifting). Good holding power, resisting withdrawal.
Duplex head	S			Where nails are to be re-drawn – formwork etc.
Cut nails				
Cut clasp	S	SC		Fixing to masonry (light weight (aerated) concrete building blocks etc.) and carpentry. Good holding properties.
Flooring brad (cut nail)	S	SC		Floor boards to joists (good holding-down qualities)
Panel pins				
Flat head	S	SC, Z		Beads and small-sectioned timber
Deep drive		G		Sheet material; plywood; hardboard
Masonry nails	S/hardened and tempered	Z		Direct driving into brickwork, masonry, concrete. (Caution: not to be driven with hardened-headed hammers. Goggles should always be used.)
Staples (mechanically driven)	S	Z	*Temporarily bonded	Plywood, Fibreboards, Plaster-boards, Insulation board or plastic films to wood battens
Corrugated fasteners ('dogs')	S	SC	Joint	Rough framing or edge-to-edge joints
Star dowel	A	SC		An alternative to hardwood dowel for pinning mortise and tenon or bridle joints

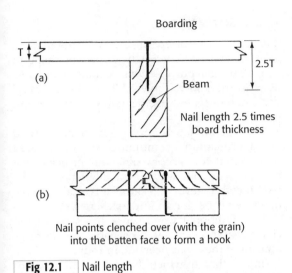

Nail length 2.5 times board thickness

Fig 12.1 Nail length

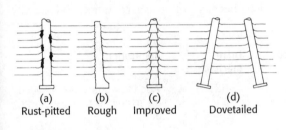

| (a) | (b) | (c) | (d) |
| Rust-pitted | Rough | Improved | Dovetailed |

Fig 12.2 Resistance to withdrawal

required to withdraw the nail, and whether the head pulls through the material. The ease or difficulty of withdrawal emphasises the importance of choosing the correct nail for the job. Apart from the nail size or type of head, resistance to withdrawal could be due to the following, shown in figure 12.2;

● type of wood,
● rust and pitting (fig. 12.2(a)),
● surface treatment of the nail, e.g. rough edged cut nail (fig. 12.2 (b)) or galvanised etc.,
● nail design, i.e. improved nails (fig. 12.2 (c)),
● dovetail nailing (fig. 12.2(d)).

Nails are more commonly associated with joints that require lateral support – preventing one piece of timber sliding on another. Nails in this instance are providing lateral resistance (see fig. 12.3), and, for this to be sustained, resistance to withdrawal is vital.

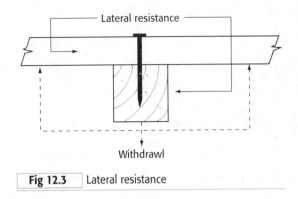

Fig 12.3 Lateral resistance

12.1.3 Nail spacing

When nailing wood, splitting can be a problem and can occur when:

● nailing too near to the edge or end of a piece of timber,
● the nail gauge is too large for the wood section (especially small sections of hardwood),
● nailing one nail behind another in-line with the grain,
● using an oversized nail punch,
● trying to straighten bent nails with a hammer.

If the above cannot be resolved by using other types or sizes or repositioning, then the following remedial measures could be considered;

● remove the nail's point – N.B. this reduces holding power,
● pre-bore the timber being fixed,
● use oval nails,
● use lost-head nails,
● remove bent nails – bent nails never follow a true course.

Figure 12.4 shows examples of minimum nail spacing when fixing to softwood – timber to timber.

12.2 Wood screws

Wood screws have a dual function – not only do they hold joints or articles together, they also act as a permanent clamp, which in most cases can be removed later for either adjustment or modification purposes.

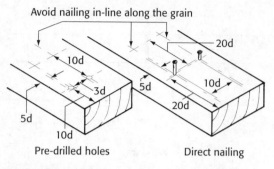

Avoid nailing in-line along the grain

10d
20d
3d
5d
10d
5d
20d
5d
10d

Pre-drilled holes   Direct nailing

d = Diameter of nail

Fig 12.4   Guide to minimum nail spacing when fixing timber to timber. Note: spacing is between nails not centres

12.2.1 Screw types

There is a vast variety of screws on the market, and knowing the correct type, size, or shape to suit a specific purpose will become a valuable asset to the joiner. Table 12.2 illustrates several wood-screw types together with their use and driving methods

Wood-screw labels serve as a quick method of identification, combining an abbreviated description with a screw silhouette or head style and a colour code which generally represents its base metal. Figure 12.5 shows a typical example of a screw label for an imperial screw size.

The traditional wood screw has seen several changes over recent years, probably the most significant of these has been the introduction of the Twinfast thread which halves the number of turns required to drive the screw home. Then came the Supafast threaded screw (see fig. 12.6), which, because of its sharp point and steeply

Table 12.2   Wood-screw fixing devices

Screw type	Material	Finish/ treatment	Head shape/ style	Drive mechanism	Application
Wood screw (traditional) (Fig. 12.7)	S, B, BR A, SS	B, BR, CP, J, N, SC, SH, Z	CKS RND RSD	SL SD	Wood to wood; metal to wood, e.g. ironmongery, hinges, locks, etc.
Twinfast wood screw	S, SS	B, SC, SH, Z	CKS, RND	SL, SD	Softwoods; particle board, fibreboard, etc. Drive quicker than conventional screws, having an extra thread per pitch for each turn.
Supafast (Supascrew and Mastascrew) (Fig. 12.6)	S	BZ	CKS	SL, SD	All hard and soft woods as well as man-made boards. Spaced thread and sharp point for quicker insertion. Hardened head and body, so virtually abuse-proof.
Coach screw	S	SC, Z	SQ	SP	Wood to wood; metal to wood (Extra-strong fixing)
Clutch screw	S, SS	SC, Z			Non-removable – ideal as a security fixing
Mirror screw	S, B	CP	CKS DM		Thin sheet material to wood – mirrors, glass, plastics
Dowel screw (double-ended) cupboard handle, etc.	S	SC			Wood to wood – Concealed fastener,
Hooks and eyes	S	B, CP, SC			Hanging – fixing wire, chain, etc.

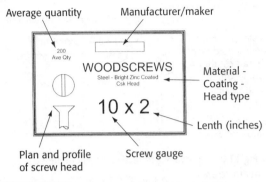

Average quantity

Manufacturer/maker

Material -
Coating -
Head type

Lenth (inches)

Plan and profile
of screw head

Screw gauge

Fig 12.5 Wood screw labelling (Imperial)

Table 12.3	Guide to imperial wood screw sizes *(with comparable metric screw shank diameter)*	
Screw gauge no.	**Metric dia (mm) -shank**	**Screw length (inches)**
4	3.0	½″
6	3.5	¾″
6	3.5	1″
7	*	1″
8	4.0	1″
8	4.0	1¼″
8	4.0	1½″
8	4.0	1¾″
8	4.0	2″
8	4.0	2½″
10	5.0	1¼″
10	5.0	1½″
10	5.0	1¾″
10	5.0	2″
10	5.0	2½″
10	5.0	3″
12	5.5	4″

Note:
When quoting screw sizes in the trade, it is common practice to state the screw length first.
* No metric equivalent.
Other sizes are available.

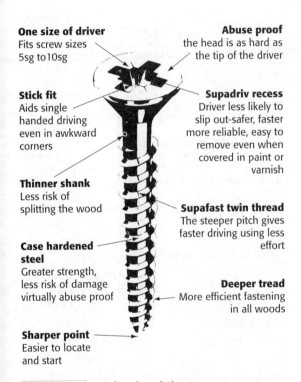

One size of driver
Fits screw sizes
5sg to 10sg

Abuse proof
the head is as hard as
the tip of the driver

Stick fit
Aids single
handed driving
even in awkward
corners

Supadriv recess
Driver less likely to
slip out-safer, faster
more reliable, easy to
remove even when
covered in paint or
varnish

Thinner shank
Less risk of
splitting the wood

Supafast twin thread
The steeper pitch gives
faster driving using less
effort

**Case hardened
steel**
Greater strength,
less risk of damage
virtually abuse proof

Deeper tread
More efficient fastening
in all woods

Sharper point
Easier to locate
and start

Fig 12.6 Supafast threaded screws

pitched thread, not only gives easy starting but drives even faster – its shank diameter is less than the diameter of the thread and therefore splitting the wood is reduced to a minimum. It is particularly useful for screwing into chipboard.

'Mastascrew' (not illustrated) with its slotted head, and 'Supascrew' (fig 12.6) with its 'Supadriv' (Posidriv) recessed head both with 'supafast' thread are hardened after manufacture. This means it is virtually impossible to

damage the screws when putting them in or taking them out because the head is as tough as the screwdriver blade or bit – particularly helpful for power driving.

12.2.2 Screw sizes

Because we are now given the choice of whether to use imperial (UK traditional sizes) or metric sizes, table 12.3 & 12.4 shows how the sizes differ. The most important aspect when fixing hardware (ironmongery) such as door hardware, is whether the hole sizes match the screw gauge, and whether the screw shank is fully or part threaded.

12.2.3 Preparing material to receive screws

Figure 12.7 shows how materials should be prepared to receive *traditional screws*, namely by boring:

● a clearance hole to suit the screw shank,
● a pilot hole for the screw thread,
● a countersink to receive the screw head – if required.

Table 12.4   Guide to metric wood screw sizes (*with comparable imperial screw guage numbers*)

Metric dia (mm) -shank	Imperial screw guage no.	Screw length (millimetres)
3.0	4	12
3.5	6	20
3.5	6	25
4.0	8	25
4.0	8	30
4.0	8	40
4.0	8	50
5.0	10	30
5.0	10	40
5.0	10	50
5.0	10	60
5.0	10	70
5.0	10	80
6.0	*	70
6.0	*	80
6.0	*	100

Note:
Other sizes are available.
* No imperial equivalent.

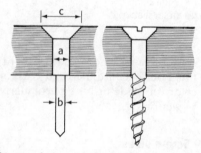

(a) Clearance hole   (b) Pilot hole   (c) Countersunk

Fig 12.7   Preparing to receive a traditional wood screw

A bradawl can be used to bore pilot holes in softwood. Failure to use pilot holes could result in the base material splitting and/or losing holding power.

Alternatively, the clearance hole, pilot hole and countersink or counterbore can be carried out in one operation (see fig. 5.52(p)).

Figure 12.8 shows how materials should be prepared to receive fully threaded screws.

12.2.4 Screw spacing

Figure 12.9 and table 12.5 show and indicate the recommended spacing of wood screws. By using

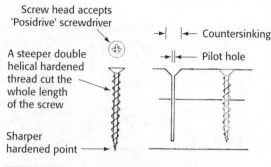

Fig 12.8   Preparing to receive a fully threaded modern screw

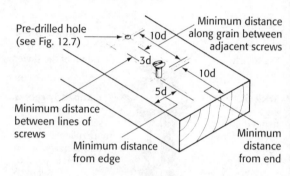

d = Diameter of wood screw shank

Fig 12.9   Minimum screw spacing when driven into pre-drilled holes

Table 12.5   Spacing of woodscrews

	Driven without Pre-drilled Holes	Driven into Pre-drilled Holes
Distance from end	20 D	10 D
Distance from edge	5 D	5 D
Distance between lines of screws	10 D	3 D
Distance along the grain between adjacent screws	20 D	10 D

D = Diameter of Woodscrew

these figures, the risk of splitting the base material is reduced, and maximum holding power is encouraged.

12.2.5 Screw cups caps and domes

Screw caps receive the countersunk head of the screw and increase the surface bearing area of the screw head – the screw eye is still visible and accessible. Caps and domes are used to totally conceal the screw head yet still provide access to the screw eye. Table 12.6. gives several examples of use and application.

12.2.6 Coach screws

The shank and tread is similar to a traditional wood screw. The head as can be seen in table 12.2 is usually square in plan, it is therefore driven in via a spanner.

Its main use is where an extra long or strong fixing is required – either wood to wood or metal to wood.

Table 12.6 Screw cups and caps

	Screw cups	Cover domes
Material	B, SS	P
Finish/colour	SC, N	Black, white, brown
Shape	CKS CKS	DM DM DM (a) (b) (c) SD
Application	Countersunk flange increases screw-head bearing area. Used where screw may be re-drawn, e.g. glass beads, access panels, etc.	(a) slots over screw; (b) slots into screw hole; (c) slots into Superdriv screw head. Neat finish yet still indicates location.

12.3 Threaded bolts

For the purpose of quick reference, table 12.7 illustrates these bolts which the carpenter and joiner are likely to encounter. But those worth mentioning separately are the:

- Coach or carriage bolt, and
- Handrail Bolt

Coach or carriage bolt (fig 12.10) – probably the most common bolt – figure 12.10(a) shows how these bolts are used to bolt together (wood to wood) wall panels of a sectional portable building. Figure 12.10(b) shows a steel band (gate and garage door hanging device) being bolted to a wood gate.

Handrail Bolt (fig 12.11) – traditionally used for end-to-end fixing handrails together, but can be applied to many other end-to-end fixings, such as bay window sills or arch door frames to name a few.

Note: *the strong hexagonal-headed bolt, together with washers and timber connectors, will be dealt with in connection with roof trusses in Book 2.*

12.4 Fixing plates

Figure 12.12 shows a few of the many fixing plates available. Some plates are multi-purpose, whereas others carry out specific functions. For example, movement plates have elongated slots to allow for either timber movement and/or

Table 12.7 Threaded bolts

Bolt type	Material	Head	Nut and bolt	Application
Hexagonal head	S			Timber to timber; steel to timber (timber connectors (etc.)
Coach bolt (carriage bolt)	S			Timber to timber; steel to timber (sectional timber buildings, gate hinges, etc.)
Roofing bolt	S, A			Metal to metal
Gutter bolt	S, A			Metal to metal
Handrail bolt	S			Timber in its length (staircase handrail, bay window sill, etc.)

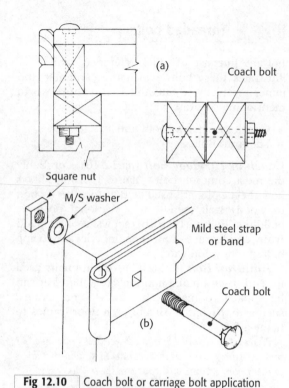

Fig 12.11   Handrail bolt application

fixing adjustment, glass plates act as a hanging medium for fixing items to walls etc. Multi-purpose plates include angles, straights, tees, etc. aid or reinforce various joints used in carcase construction. Application of corner plates is shown in figure 12.13.

Anchor plates Straps and Hangers (fig 12.14) – There are also many different forms of metal straps, framing anchors, and joist hangers etc. that in the main are designed for a specific purpose. When used for joining components as part of a structural design their fixing is strictly controlled with regard to placement and the type, size and gauge of fixings (nails).

Fig 12.10   Coach bolt or carriage bolt application

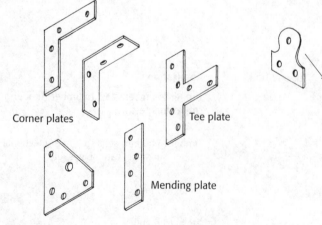

Fig 12.12   Metal fixing plates

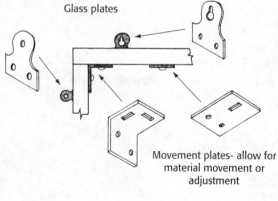

Chemical cartridge applicator

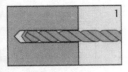

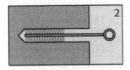

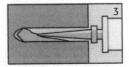

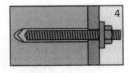

Drill hole to required
depth and diameter
to suit fixing

Thoroughly remove debris
from within the hole

Insert nozzle of cartridge
and compound into hole

Insert fixing with twisting
motion - alow to set

Fig 12.30 | Injection fixing

Cartridge type suitable for installation into all types of masonry (fig 12.30). This system can consist of a chemical capsule containing resin and a phial of hardener, which is inserted into a pre-bored hole. When the fixing is applied, the capsule breaks, the resulting mixture is chemically activated (begin the hardening process) securing the fixing into the hole. A curing (setting) time must be allowed for before any attachments can be made.

Alternatively as shown in figure 12.30, a two part mix can be injected into the hole using a cartridge gun before inserting the fixing (threaded stud or bolt).

Practical Projects

The first of these projects known as the 'Porterbox' system has been designed as a simple alternative to traditional methods of storing hand tools, power tools, accessories and hardware (ironmongery).

The system can be simply adapted to suit individual needs; for example, the type of tools and equipment and the place of work, be it a workshop or building site.

The main component (project) of the system is the *Porterbox* drop-fronted tool box. It has three optional add-on attachments:

- *Portercaddy.*
- *Portercase.*
- *Porterdolly.*

These can then be backed up with the *Porterchest* – originally designed to accommodate items of hardware, but just as at home holding an assortment of hand tools or even portable powered hand tools – the choice is yours.

Finally, the *Joiner's work bench and Carpenter's saw stool.* Much has been written about the most suitable bench and many proprietary models are available on the market. What I have tried to do here, is to design a type of bench that can be easily dismantled, yet still keeps the general principal of layout and size.

On the other hand, carpenters saw stools which are probably more commonly known as 'trestles' and have used the same method of construction for many years.

These practical projects (activities) have been itemised as follows:

1 Porterbox
2 Portercaddy
3 Portercase
4 Porterdolly
5 Porterchest
6 Porterbench
7 Portertrestle

Practical project 1: Porterbox – drop-fronted tool box and saw stool

This design is a practical alternative to the traditional joiner's tool box previously mentioned. The main idea behind this new concept was to allow all the established practices, such as leaving the front flap open for tool access during the work period, using the box top as a work surface and resting place during break periods etc., to continue, but in a safer and more practical way. This meant that:

1 The front flap would not project more than is necessary beyond the base of the box whilst still allowing easy access to the tools inside, thereby reducing the area of floor space taken up by the flap and the risk of passers-by tripping over it.
2 The box top would be capable of being used as an unobstructed flat work surface without the interference of a handle or latching mechanism.
3 Easy and quick access to racked tools would be available.

Other factors built into the design also meant that:

4 Construction would be strong without being too heavy.
5 It would be large enough to accommodate the basic tool kit together with a selection of special tools.
6 A single means of securely locking both top

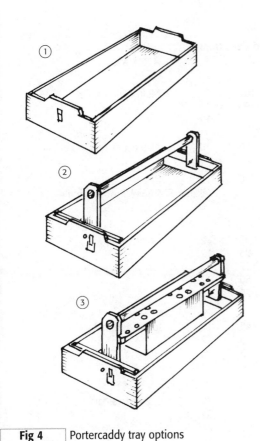

| **Fig 4** | Portercaddy tray options |

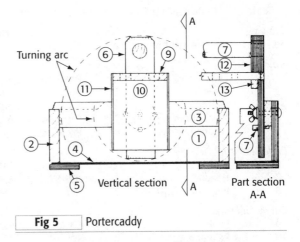

| **Fig 5** | Portercaddy |

been made for a detachable tool rack to be incorporated in the construction – when not in use the detachable rack can be used on top of the workbench. The folding bar carrying handle is retained within the tray when attached to the toolbox.

The *Portercaddy is* attached and detached by the use of heavy-duty quick release toggles, or heavy-duty trunk case clasps.

Sectional details of construction are shown in figure 5. A full cutting list is given in table 2, in which components are named and numbered to coincide with the drawing.

Method and sequence of construction
(fig. 6)

Rod

Using the rod used to construct your *Porterbox,* extend the sections to accommodate all the attachments, taking extra care when allowing for

clearance about the turning circles for the bar handle and tool rack as indicated in figure 5. This will permit both handle and tool rack to rotate freely through 180°.

Note: time spent at this stage of the proceedings usually means time and materials saved later on.

Tray

Tray ends [1] (12 mm Plywood) – cut to shape (allow for keyed upstand). Then, while temporarily tacked together (face sides outermost), plane face edges, dress ends square and form the toolbox locating key upstands.

Tray sides [2] – cut to length – check face sides (outside faces) for twist – remove any twist by planing. Temporarily tack together (face sides outermost) and plane face edges. Finally dress ends square and true to length – separate sides.

Corner joints – double-through or lapped dovetailed joints are recommended. Dovetails should be formed in the plywood ends [1]. When cut, the ends can be separated. Progress to marking and cutting pins in the side's [2].

Keyway – using the formed locating keys [l] as a template, carefully mark out and label the matching keyways onto the base of the Porterbox ends. Cut out the keyways, allowing only the very minimum of play. It is advisable to code these ends so that they match when the tray is assembled.

Bearers' [3] – prepare these to be in line with the top of locating keyed upstands (12 min above tray ends), and notched over both sides.

| Table 2 | Cutting list for Portercaddy, Portercase, and Porterdolly |

Drawing no.	Item description	No. off	Material	Finished sizes (mm)			Remarks
				L	W	T	
Tray							
1	End	2	Plywood	300	107	12	Multi-ply-form key
2	Sides	2	SW	676	95	18	
3	Bearers	2	SW	300	46	12	
4	Tray bottom	1	Plywood	676	315	4	
5	Locating blocks	4	Plywood	75	50	10	Also act as feet
Caddy tool rack							
6	Hanger	2	Plywood	215	50	12	
7	Handle	1	HW/SW	630		25 dia	Dowel/Broom handle
8	Peg – anti swivel	2	HW	50		6ϕ	Dowel
Caddy tool rack							
9	Shelf	1	SW	626	95	12	Holes to suit
10	Pelmet ends	2	SW	125	95	12	
11	Pelmet face	2	PL/Hdbd	300	137	3	
12	Hanger cleat	2	Plywood	64	50	12	
13	Shelf bearer	2	SW	50	22	12	
Case lid							
14	Frame	2	SW	628	45	12	Joints optional
15	Frame	2	SW	300	45	12	
16	Lock block	1	SW	75	45	12	
17	Top	1	PLY/Hdbd	628	270	3	
Dolly							
18	Bearer	2	SW	430	70	25	Support castors
19(a)	Rails (solid)	2	SW	540	45	34	
19(b)	Rails (laminated)	2	SW	540	45	25	
		2	SW	540	45	12	
20	Locating strip	2	SW	215	12	25	
Hardware							
21	Coachbolts	2		50		69	Washer/wingnut (2)
22	Toggles*	2					
23	Hinges (cranked)	2/3					Cabinet hinges, non-recessed type
24	Cupboard lock	1					
25	Case handle	1					
26	Castors – heavy duty	4					Base plate type (locking)

Key: SW – Softwood eg. redwood, whitewood or similar species ($\pm$ 1 mm). HW – Hardwood; straight grained species ($\pm$ 1 mm). PLY – Plywood; exterior quality not necessary. Hdbd – Hardboard (good quality); less expensive substitute for plywood. * – Stong lockable case trunk fasteners (clasp) can be used.

Tray bottom [4] (4 min Plywood) – cut to size. Plane two edges to form one square corner. Location blocks [5] (10 min Plywood) – cut and plane to size.

Tray assembly:

1 Sand all inside faces of ends, sides, and bottom.
2 Glue and screw bearers in line with top of keyed upstand to plywood ends.
3 Glue and make all corner joints (cramp as necessary). Square framework by positioning square corner of plywood over frame. Glue underside of tray, tack bottom in position.

Note: At this stage, position the end keys into the box keyways, and check that the tray is square with it – adjust as necessary.

4 When alignment is true, screw (1″ × 8 Csk) Bottom [4] onto sides and ends.

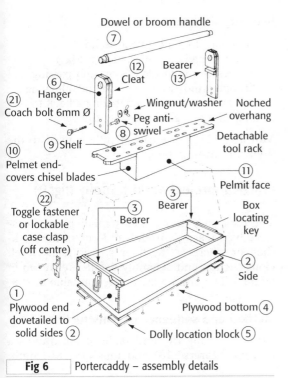

Dowel or broom handle
⑦

⑫ Cleat   Bearer ⑬

⑥
㉑ Hanger
Coach bolt 6mm Ø

Wingnut/washer   Noched
Peg anti-   overhang
⑧ swivel

⑨ Shelf   Detachable
tool rack

⑩
Pelmet end-
covers chisel blades

③ Bearer   Box
㉒   ③   locating
Toggle fastener   Bearer   key
or lockable
case clasp
(off centre)

②
Side

①
Plywood end
dovetailed to   Plywood bottom ④
solid sides ②
Dolly location block ⑤

Pelmit face ⑪

| **Fig 6** | Portercaddy – assembly details |

5. When set, dress all joints and trim bottom all round.
6. Attach Porterdolly locating blocks to each corner with glue and screws (1″ × 8 Csk).

Caddy

Hangers' [6](12 mm Plywood) – cut to size, tack together, then plane to shape. Bore a series of holes to receive

1 shouldered handle (25 mm Dia),
2 coach bolt (6 min Dia),
3 anti-swivel and locking peg (6 mm Dia).

Handle [7] – cut dowel to length. Reduce diameter of both ends to 22 mm by forming shoulder 12 mm in from each end.

Pegs [8] – prepare from 6 mm and 12 mm dowel.

Assembly – transfer, and bore holes through and/or into bearer and ends to take coach bolt and peg holes. Glue handle to hanger. Position coach bolt with washer and wing-nut on the inside of tray. Anti-swivel pegs should be a push fit.

Caddy tool rack
Shelf [9] – cut to length. Prepare by cutting notches for hangers and holes for tools.

Pelmet ends [10] – cut to length and square ends. Glue and nail to underside of shelf.

Pelmet face [11] (3 mm Plywood) – cut to size. Glue in squarely to pelmet ends and shelf, trim the edges and sand the whole assembly.

Hanger cleats [12] (12 mm Plywood) – cut to size and tack together, shape and bore holes in-line with hangers. Glue and nail one to each hanger.

Shelf bearers [13] – cut to length – glue and screw (1″ × 6 Csk) to the hangers. Dress ends square with hanger edges.
  Note: When a detachable caddy tool rack is to be an option, hanger cleats are used. The bar handle is left loose to rotate within the widened hanger.

Attachment
The tray is held to the box by two toggles [22] (one at each end) or heavy-duty trunk case clasps. If security is important, types with a locking eye are available – alternatively a small hasp and staple could be used.

Practical project 3: Portercase (fig. 7)

By using the same tray construction as the *Portercaddy* with the addition of a hinged lid (fixed or detachable), the base unit can be transformed into a useful toolcase as shown in figure 7. An additional cutting list is shown in table 2.

Method and sequence of construction (fig. 8)
Rod – as the tray, but with the addition of the lid.

Lid framework – choose a suitable corner joint for members [14] and [15]. For example:

a mortise and tenon (through or stopped),
b halving joint,
c dowelled,
d bridle.

A butt joint could be used, with a triangular plywood gusset glued and pinned over the underside of each corner. Alternatively, a strip of

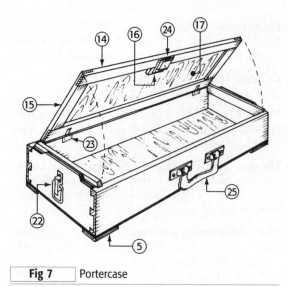

Fig 7 | Portercase

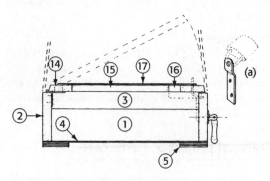

Fig 8 | Portercase – section details

plywood fixed along the full length of the hinged side of the lid would then leave a useful pocket within the underside of the lid.

Note: before gluing joints together do not forget to clean-up the inside edges, and to check that framework is square with the base tray. After the joints are made, dress and sand the faces before applying the top.

Top [17] – cut to size, leaving a margin all round. Plane and bullnose edges. Pin to pre-glued upper face of framework.

Hinging lid [23] – use three 50 mm non-recess cranked cabinet hinges.

Note: if the case is to be used in conjunction with the caddy, then consider using the lift-off type of cranked hinge.

Security – alternatives include:

- Cupboard lock [24] fixed to lid.
- Purpose-made locking bar (fig. 8(a)) – screwed to the inside of box, and passing through a mortise hole in the lid, allowing enough projection to enable a padlock to be inserted through the hole in the bar.

Mobility [25] – the case handle is fixed to the side with through-bolts.

Practical project 4: Porterdolly (fig. 9)

The *Porterdolly* provides easy mobility via a cas-tored baseboard – a useful 'add-on' for the work-shop or wherever the floor is reasonably level. Construction and assembly details are shown in figure 9. The cutting list is given in table 2.

Method and sequence of construction

Frame – using the underside of the tray as a rod, cut bearers (18) and rails (19) to length.

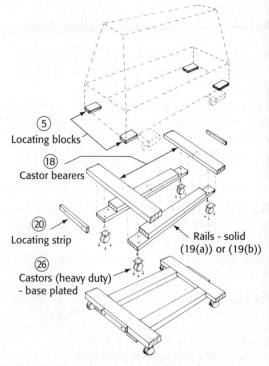

(5) Locating blocks

(18) Castor bearers

(20) Locating strip

Rails - solid (19(a)) or (19(b))

(26) Castors (heavy duty) - base plated

Fig 9 | Porterdolly – assembly details

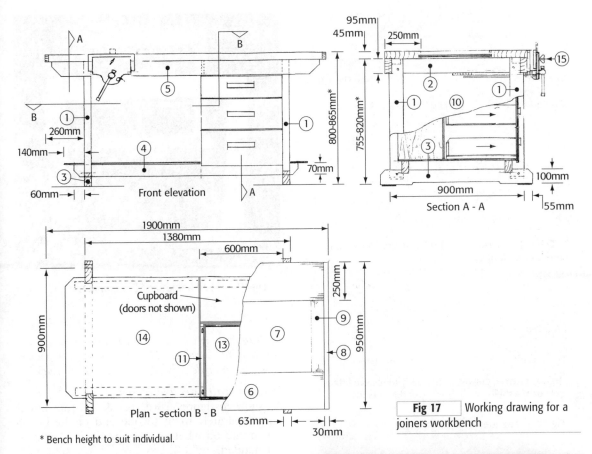

Note: *when marking the position of housings, the facings [5] should be marked at the same time as the runners [4].*

Facings (fig 18c & 20) – constructed from 50 × 100 (Redwood), facings [5] are housed 12 mm deep to receive outer edge of the notched stiles. A stopped housing 12 mm deep to receive a cross bearer (mid support 25 × 100 for the bench top) should be formed at the same time.

Before fixing underside corners at both ends should as shown in figure 17 bevelled.

Note: *Joints should be a tight fit, and secured either by screwing through the face into the end frames, or alternatively as shown in figure 21 by using heavy-duty angle brackets.*

Bench top

The top is made-up into a frame – consisting of:

Two planks [6] (50 × 250) – preferably Beech (HW), alternatively, a hard softwood such as Douglas fir (SW) may be used. A plough groove is cut into the inner edges to receive the central

panel [7]. This groove can also be cut into the plank ends to receive end trims with a loose tongue (this would help to stabilise any potential end grain distortion.

Central panel [7] – plywood (12 mm) its length will be 20 mm longer than the planks (10 mm left at either end to fit into end trims [8]).

(a) Completed bench without vice fitted

Fig 17 Working drawing for a joiners workbench

* Bench height to suit individual.

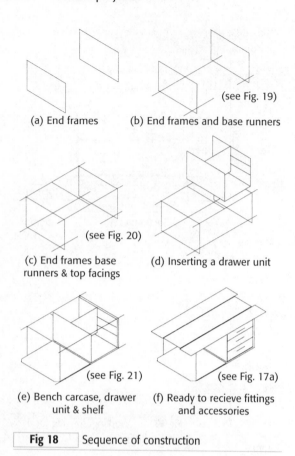

(a) End frames

(b) End frames and base runners (see Fig. 19)

(c) End frames base runners & top facings (see Fig. 20)

(d) Inserting a drawer unit

(e) Bench carcase, drawer unit & shelf (see Fig. 21)

(f) Ready to recieve fittings and accessories (see Fig. 17a)

Fig 18   Sequence of construction

End trims (25 × 50) – same material as planks. A plough groove is cut into the inner edges to receive the central panel and if used a loose tongue. Alternatively a stopped groove can be cut just long enough to receive the panel. The portion that covers the plank end grain can then either be glued and screwed/dowelled into the plank ends.

Don't forget the importance of covering the end grain of planks, or large solid wood boards if distortion is to be avoided. Potentially this is where the greatest amount of any moisture intake into the wood will occur.

Bevelled fillets [9] (optional) – cut to suit. Traditionally these are feathered into the bench-well to provide an easy means of sweeping out the well. *Today the technique of 'sweeping out' must be regarded as a potentially dangerous bad practise.* Such operations will inevitably result in waste particles of various size being freely discharged into the air.

Fig 19   End frames and base runners

Note: *the top will be attached to the bench carcase via steel angle brackets. This will enable easy detachment, and provide easy access to vices mountings.*

Drawer unit carcase *(optional)*
As show in figures 18(d) (e) & 21, the unit is built separately to fit to one end of the bench depending on whether the operative is right or left-handed.

The unit carcase consists of:

Carcase side panel's [10] plywood (12 mm) – proprietary drawer runners [11] are fixed to the sides, and grooves are cut to house the mid-panel ends and *base board.

Note: *side panels should be allowed to extend below the plywood carcase base.*

Carcase mid-panel – plywood (12 mm) – size to suit length of drawer fronts with provision for the housings.

Carcase base – plywood (10 mm) – width as the mid-panel, and length as the side panels.

Drawers
As shown in figure 21 drawers for this prototype model were made-up using proprietary combined metal drawer sides [12] and runners. Drawer fronts and backs in this case were made from 12 mm plywood screwed to adjustable fixing brackets attached to the drawer sides. The 6 mm plywood drawer bottoms [13] sit on a

Fig 20 End frames, base runners, top facings and cross bearer

metal lip formed into the drawer sides, and is fitted into drawer fronts and glued and pinned to the under side of the drawer backs.

Note: several methods of drawer construction are featured in book 3 under the heading of joinery fitments.

Shelves
Shelves – plywood (10–12 mm) – as shown in figures 17 & 18(e) with the exception of the shelves within the cupboard behind the drawer, others can extend outside the end frames. In the case of the large shelf [14] cross support bearers may be required.

Bench assembly
Figures 17(a) & 18(f) show the main components assembled ready to receive a bench vice [15] or vices, and other bench fixtures.

Bench fixtures
For the purpose of this chapter bench fixtures will include:

● Joiners bench vice (with or without a cramping dog)
● Bench-end vice with cramping dogs
● Bench stops
● Cramping (clamping) bench dogs
● Bench holdfast

Joiner's bench vice – figures 17[15] & 24 shows a vice that incorporates a quick release mechanism (regarded as essential for the bench hand)

Fig 21 Bench carcase, drawer unit and shelf

with a height adjustable cramping dog, which is used in conjunction with bench top inset dogs.

Bench-end vice with cramping dogs – figure 24 shows an end vice which incorporate two height adjustable cramping dogs. These vices usually have a steel movement mechanism with full hardwood jaws. They are used in conjunction with a row or rows of bench dogs.

Bench stops – these are sited towards the ends of the bench. Three examples are shown in figures 22 and 23.

Purpose made bench stops (fig 22) – sizes vary to suit bench top thickness and individual requirements, for example the one shown would be

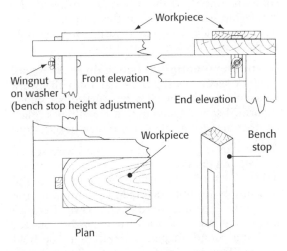

Fig 22 Purpose made bench stop. NB. Bench facings not shown)

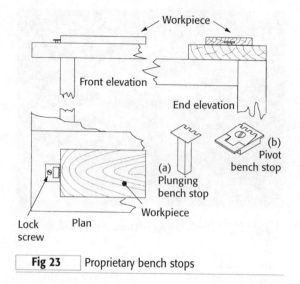

Fig 23 | Proprietary bench stops

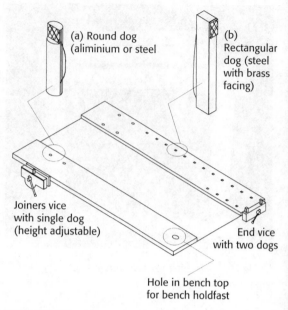

Fig 24 | Metal bench dogs

made from 22 × 40 mm section Beech and 150 mm long. The wing nut and washer hold the stop to the required height – when not in use the stop is lowered in line with the bench top.

Proprietary metal bench stops – figure 23(a) shows a plunging steel bench stop with serrated teeth to grip the workpiece when pushed up against it. It fits into a metal sleeve, which is mortised into the bench top. It is locked to the required height by means of a set-screw accessed from the top face plate. Figure 23(b) shows another type that again fits flush with the bench top when closed. In this case it is made from a soft alloy. Height adjustment via a hole in the spring loaded serrated stop.

Cramping (clamping) bench dogs – these are used to secure workpiece's flat onto the work bench to enable them to be worked on safely. Figure 24 shows two different metal types (plastic dogs are also available) and their application.

Round dogs (fig 20a) – with integral wire spring to help retain them within an 18 mm round hole. Holes can be drilled at intervals in the bench top to suit increments relevant to the maximum opening of the vice jaws. Because these dogs are cylindrical they can be rotated, enabling round shaped workpieces to be held secure. The *knurled* face of the dog may be slightly undercut to help hold the workpiece down to the bench top.

Rectangular (Steel) dogs (fig 20b) – with an integral flat spring to help retained them within a rectangular hole 19 × 25 mm. Holes are mortised into the bench top at intervals to suit increments relevant to the maximum opening of the vice jaws. These dogs can be as long as 175 mm, therefore the thickness of the bench top will reflect the amount that the dog can project above its surface whilst still being held secure within its mortise hole.

Bench holdfast – one common type which fits into a metal sleeved hole cut into the bench top (usually to the opposite end to the main vice) is, together with its application, shown in section 5.88 figure 5.89. Other types are available, one can be used in conjunction with holes drilled for round dogs.

Note: Some holdfasts can also be used horizontally into holes cut into reinforced bench facings or legs, to hold or support the other end of vice-held boards/sheets etc.

Practical project 7: Portertrestle – traditional saw stool (fig. 25)

As shown in figure 25(a) & (b) this design uses traditional methods of construction. Trestles are used individually or in pairs.

Index